Medullary Thyroid Carcinoma

Beiträge zur Onkologie
Contributions to Oncology

Vol. 17

Series Editors
S. Eckhardt, Budapest; *J.H. Holzner,* Wien; *G.A. Nagel,* Göttingen

S. Karger · Basel · München · Paris · London · New York · Tokyo · Sydney

Medullary Thyroid Carcinoma

Leonard J. Deftos, MD
Professor of Medicine
University of California, San Diego
Chief of Endocrinology
San Diego VA Medical Center
La Jolla, Calif., USA

5 figures and 2 tables, 1983

S. Karger · Basel · München · Paris · London · New York · Tokyo · Sydney

Beiträge zur Onkologie
Contributions to Oncology

National Library of Medicine, Cataloging in Publication
  Deftos, Leonard J.
  Medullary thyroid carcinoma/
  Leonard J. Deftos. – – Basel; New York: Karger, 1983.
  (Contributions to oncology = Beiträge zur Onkologie; v. 17)
  1. Carcinoma 2. Thyroid Neoplasms I. Title II. Series: Beiträge zur Onkologie; v. 17
  W1 BE444N v. 17 [WK 270 D316m]
  ISBN 3-8055-3703-4

Drug Dosage
  The authors and the publisher have exerted every effort to ensure that drug selection and dosage set forth in this text are in accord with current recommendations and practice at the time of publication. However, in view of ongoing research, changes in government regulations, and the constant flow of information relating to drug therapy and drug reactions, the reader is urged to check the package insert for each drug for any change in indications and dosage and for added warnings and precautions. This is particularly important when the recommended agent is a new and/or infrequently employed drug.

All rights reserved.
  No part of this publication may be translated into other languages, reproduced or utilized in any form or by any means, electronic or mechanical, including photocopying, recording, micro-copying, or by any information storage and retrieval system, without permission in writing from the publisher.

© Copyright 1983 by S. Karger AG, P.O. Box, CH–4009 Basel (Switzerland)
  Printed in Switzerland by Thür AG Offsetdruck, Pratteln
  ISBN 3-8055-3703-4

Contents

Acknowledgements VIII
Dedication IX

History 1
Histogenesis 4
Incidence 5
Genetics 7
Pathology 8
    Introduction 8
    Gross 8
    Histology 9
    Amyloid 12
    Tumor Calcification 13
    C-Cell Hyperplasia (CCH) 14
    C-Cell Adenoma 15
Natural History 16
Calcitonin 18
    Biochemistry 18
    Calcitonin Secretion by Normal C-Cells 20
        Calcium and Related Minerals 20
        Gastrointestinal Factors 21
        Neuroendocrine Factors 21
        Age and Sex 22
        Other Factors 22
    Secretion by MTC 23
        Provocative Testing 24
            Glucagon 24
            Calcium 24
            Pentagastrin 24
            Pentagastrin versus Calcium 25
            Whiskey 26
            Magnesium 26
        Venous Catheterization Procedures 26
        Serial Calcitonin Measurements in the Evaluation of Therapy 27
    Immunochemical Heterogeneity 27
    Metabolism 28
    Clinical Uses of Calcitonin 29

Contents

Differential Diagnosis of Hypercalcitoninemia	30
Hypercalcitoninemia Associated with Malignancy	30
Ectopic Calcitonin Production	30
Eutopic (Thyroidal) Calcitonin Production	31
Artifactual Elevations in Plasma Calcitonin	32
Other Hypercalcitoninemic States	32
Hypercalcemia and Primary Hyperparathyroidism	32
Hypocalcemia	33
Renal Disease	33
Bone Disease	34
Pancreatitis	34
Production of Other Bioactive Substances by MTC	36
ACTH Production and Cushing's Syndrome	36
History	36
Pathogenesis	37
Clinical Features	38
Histaminase	38
Prostaglandins	39
Dopa Decarboxylase and Biogenic Amines	40
Serotonin and Related Factors	41
Carcinoembryonic Antigen (CEA)	41
Nerve-Growth Factor	41
Other Peptides	41
Other Substances	42
Clinical Consequences of Hypercalcitoninemia	43
Biological Effects of Calcitonin	43
Bone	43
Kidney	43
Gastrointestinal Absorption of Calcium	43
Other Effects	44
Role of Calcitonin in Mineral Metabolism	44
Newer Concepts – Calcitonin as a Neurotransmitter	46
The Patient with MTC	47
Bone Disease	48
Blood and Urinary Minerals	49
Kidney Stones	50
Diarrhea	50
Peptic Ulcer Disease	51
Carcinoid Syndrome	52
Hypertension	53
Pigmentation	53
Gynecomastia	53
Marfanoid Habitus	53
Mucosal Neuromas	54
Cushing's Syndrome	54
Cervical Enlargement	54

Contents

Multiple Endocrine Neoplasia (MEN)	55
Pheochromocytoma	56
History	56
Incidence	57
Clinical Manifestations	57
Diagnosis	58
Adrenal Medullary Hyperplasia	59
Hyperparathyroidism	60
History	60
Incidence	61
Pathology	61
Functional Activity of the Parathyroid Gland in Patients with MTC	63
Relationship to MTC	64
Mucosal Neuroma Syndrome	65
History	65
Pathology	67
Mucosal Neuromas	68
Oral Cavity	68
Ocular Abnormalities	68
Cutaneous Lesions	69
Gastrointestinal Abnormalities	71
Other Neural Abnormalities	72
Marfanoid Habitus	72
Other Musculoskeletal Abnormalities	72
Other Locations of Neuromas and Other Abnormalities of the Oral Cavity	73
Clinical Implications	74
Radiological Findings	74
Treatment of Multiple Endocrine Neoplasia	76
Medullary Thyroid Carcinoma	76
Pheochromocytoma	77
Neuromas	78
Hyperparathyroidism	78
References	80
Subject Index	111

Acknowledgements

This work was supported by the American Cancer Society.

Dr. *Louis Avioli* encouraged the preparation of this monograph. *Robert Cowan, Mercedes Gacad,* and *Susan Murphy* provided assistance.

Dedication

ΑΦΙΕΡΩΜΕΝΟ ΣΤΟΝ ΕΑΥΤΟ ΜΟΥ

History

Medullary thyroid carcinoma (MTC) is a tumor of the calcitonin-producing cells (C cells) of the thyroid gland. In contrast to the follicular cells of the thyroid gland which metabolize iodine and produce the thyroid hormones T3 and T4, the significance of the C cells of the mammalian thyroid gland has been only recently established. The presence of such a population of cells was first suggested in 1876 during studies of the anatomy of the dog thyroid by *Baber* [40] who described these cells as parenchymatous. In more definitive studies in 1932, also in the dog thyroid, *Nonidez* [382] described these cells as parafollicular because of their location adjacent to the thyroid follicle. This terminology has been revived for current usage.

This distinct population of cells received little attention until the discovery of their hormonal product, (thyro)calcitonin, in the early 1960s. In perfusion studies of the thyroid-parathyroid apparatus of the dog, *Copp* et al. [141] discovered a hypocalcemic factor that caused a greater decline in blood calcium than that produced by parathyroidectomy. They proposed the existence of a hypocalcemic hormone that was secreted by the parathyroid gland and named it calcitonin. *Hirsch* et al. [257] also postulated the presence of a new hypocalcemic factor as a result of their careful studies of blood calcium in rats. They observed that surgical parathyroidectomy produced a lesser fall in blood calcium than did parathyroid ablation by electrocautery. They concluded that cautery of the adjacent thyroid gland released a calcium-lowering factor from the thyroid gland. They extracted such a hypocalcemic factor from the thyroid gland and it was named thyrocalcitonin. *Kumar* et al. [313] provided further evidence for this hypocalcemic factor with perfusion studies in the dog, but they could not distinguish between a thyroid or a parathyroid source for it. Although further evidence for a parathyroid source for the newly discovered hormone was provided by experiments in sheep [106], this issue was settled by *Foster* et al. [211] in 1964. They studied the goat, which has a separate blood supply to one set of its parathyroid glands. By perfusing the parathyroids separately they could not demonstrate a hypocalcemic factor from these glands. By contrast, perfusion of the thyroid (which also contained some parathyroid tissue) did release a hypocalcemic factor. *Foster* et al. [212] further concluded from studies in the dog that the source of calcitonin was a cell population corresponding to the parafollicular cells of *Nonidez* [382] and the parenchymatous cells of *Baber* [40]. Further histochemical and ultrastructural studies [89, 97, 207, 391, 393, 406, 407] and more specific immunohistological studies [95, 96, 346, 547] have established that this discrete population of cells – the C cells – is the source of calcitonin in man and other mammals. Despite studies to the contrary not based on specific immunochemical techniques [222], there is no convincing evidence by immunohistology of the presence of calcitonin in the thymus or parathyroid gland of man [287, 346]. The general issue of extrathyroidal calcitonin is discussed subsequently. In contrast to the situation in mammals, calcitonin in submammals is located in a distinct organ, the ultimobranchial (UB) gland [143].

In 1966, *Williams* [536] suggested that the C cells might be the cell of origin of MTC. This tumor had been recognized as a distinct pathological entity which could be distinguished from other thyroid tumors by *Hazard* et al. [242] who, in 1959, identified 21 cases of this tumor from 600 cases of thyroid cancer; because of its solid appearance and its resemblance to mammary carcinoma, he called it medullary, meaning 'solid' [243]. It is likely, however, that the previous publications [88, 93, 263, 281, 316, 491] were describing the same tumor. *Williams'* hypothesis was proven correct in 1968 when several investigators demonstrated, by bioassay, the presence of abnormal calcitonin concentrations in the plasma and tumor of patients with MTC [150, 348, 359, 362]. These findings were confirmed by many other studies. In fact, the high concentrations of this hormone in an MTC permitted the isolation, purification, and sequencing of human calcitonin [378, 463]. These biochemical and biological findings were confirmed by specific radioimmunoassays of human calcitonin in tumor and blood [128] and by histochemical and immunohistochemical studies [96, 287, 310]. Although the demonstration of C cells by anticalcitonin antisera was first reported by *Bussolati* et al. [96], these studies used antisera to porcine calcitonin which do not usually cross-react with human calcitonin and which gave negative results in the studies of *Kracht* et al. [310]. Because of this, subsequent studies using antisera to human calcitonin provide more convincing data that the cells of MTC contain calcitonin [288, 546]. These observations in humans were supplemented by the finding of tumors of the same cell population in rats, bovines, and canines [155, 312, 319, 404]. The tumor was also subsequently shown to produce a wide variety of bioactive substances.

The unique histological and biochemical features of MTC were soon embellished by its unique clinical characteristics of being commonly associated with other endocrine and nonendocrine neoplasms. A series of careful clinical studies established the association between MTC and pheochromocytomas, parathyroid neoplasms, and multiple mucosal neuromata (MMN) [152, 476, 488, 538]. Most, if not all, of these associated neoplasms share with MTC a common embryological origin – the neural crest [525].

These observations established the existence of a multiple endocrine neoplasia (MEN) distinct from MEN Type I.

Histogenesis

It is now generally accepted that the thyroid C cells, which become neoplastic in MTC and C-cell hyperplasia (CCH), are neural crest in origin [544]. This was demonstrated independently by *LeDourain and LeLievre* [320], *Pearse and Polak* [409], and *Polak* et al. [417] with different techniques. The latter demonstrated the neural-crest origin of thyroidal C cells by injecting pregnant mice with amine precursor. Cell in the fetus avid for the substances migrated from the neural crest to the last branchial pouch and ultimately to the thyroid. The former group transplanted segments of neural crest from embryonic Japanese quail into the neural-crest area of embryonic chicks. The distinct histological appearance of the transplanted cells allowed their migration to be traced from the neural crest to the UB pouch of the recipient birds. The UB pouch in birds, fish, and amphibians develops into a distinct gland, the UB gland, and is the primary residence of the C cells in these animals [168, 185]. In mammals, these cells migrate into the thyroid gland and perhaps, as will be discussed, to other sites such as the pituitary, pancreas, lung, and gastrointestinal tract [64, 174, 408]. The neural-crest origin of C cells not only offers an explanation for the association of MTC with other tumors of neural-crest origin, but also explains the histological similarities of some of these tumors and their production of a wide variety of bioactive substances [8, 417].

Incidence

MTC is not a common thyroid tumor. In most series, its incidence among thyroid cancers ranges from approximately 3 to 12% [127, 254, 270, 271, 302, 333, 384, 385, 429]. However, its clinical manifestations and functional characteristics give it more importance and interest than its incidence might warrant. The important clinical features of the tumor are its familial incidence with an autosomal dominant pattern and its association with other endocrine as well as nonendocrine tumors. The important functional characteristics of the tumor include its production of a variety of bioactive substances, the most notable being calcitonin. When these factors are considered, along with the potential for early diagnosis and surgical cure of MTC, the intensity with which this tumor has been studied can be appreciated.

Although currently reported to account for a minority of thyroid cancers, it is very likely that the reported incidence of MTC will continue to increase. It has only been recognized as a distinct clinical entity since 1967 and the diagnosis continues to be either missed or made retrospectively [333]. Recording familial distribution and establishing screening procedures for detection of the tumor will undoubtedly lead to a continual increase in diagnosis [127].

The majority of cases of MTC reported in the literature are sporadic [254, 302, 534]. However, a family history is often inadequate in establishing familial disease and more thorough evaluation often reveals the presence of familial tumor in a case originally diagnosed as sporadic [387]. Furthermore, an appreciation of the possibility of familial incidence and early diagnosis has led to increased reporting of familial cases of the tumor [127, 242, 387]. Therefore, it is likely that familial cases will represent an increasing fraction of patients with MTC.

The ratio of female-to-male involvement seems to be closer to unity than in other thyroid tumors when most large series are considered. In 73 cases of MTC out of 777 cases of thyroid cancer studied by *Hill* et al. [254], the female-to-male ratio was 1.4 to 1; this compared to 2.5 to 1 for other thyroid tumors. MTC represented 12.6% of thyroid cancer in males and 7.5% of thyroid cancer in females. The relatively higher incidence in males of this thyroid tumor probably reflects, in part, the autosomal dominant transmission of the familial form which, in contrast to the greater female incidence of other thyroid tumors, has no sex preference.

MTC can occur over a wide age span. In one large series the age at diagnosis ranged from 15 to 82 years with an average of approximately 47 years. The age at diagnosis of familial MTC is earlier than at diagnosis of sporadic, as would be anticipated. In a series of familial cases, the mean age at diagnosis was 19 years [303], whereas in a presumably sporadic series [270] the mean age at diagnosis was 53 years. In another series [127], peak incidence of familial cases was in the second decade, whereas the peak incidence of all cases was in the sixth decade. In yet another series of primarily familial cases, the diagnosis was made in the age range of 20–29 years. Several early familial cases have been recorded in the literature [200]. The tumor has been observed in a 2-year-old child [490] and in a 4-year-old child [45] with regional lymph node involvement. The youngest reported patient with MTC was 14 months of age [195]. The occurrence of MTC in a familial pattern and at an early age emphasizes the need for early diagnosis and screening procedures in the relatives of affected patients [221].

Genetics

The pattern of inheritance in essentially all series of familial MTC fulfills the criteria for an autosomal dominant characteristic although there is some evidence for recessive pattern [298, 386, 387]. The tumor is transmitted from generation to generation from a parent of either sex to a child of either sex with a high degree of penetrance [11]. The sex ratio is approximately 1 and the incidence is approximately 50%. With some exceptions, chromosomal studies in patients with MTC are essentially normal [209, 303]. Clinical studies of MTC are compatible with the two-mutational-event theory of the initiation of cancer postulated by *Knudson and Strong* [306]. According to this view, hyperplasia is believed to be the expression of the genetic mutation, which requires a subsequent somatic mutation to transform the initially mutated cell into a cancer cell [58, 278].

Pathology

Introduction

Now that MTC is appreciated to be a distinct pathological entity, most experienced pathologists can readily differentiate between this tumor and other, more common thyroid tumors [543]. However, an atypical MTC may still be mistaken for another type of thyroid tumor. More commonly, a medullary thyroid tumor classified as another type before the feature of MTC was well appreciated is correctly reclassified upon re-examination [385]. The proper identification of MTC, even retrospectively, has important implications for the patient and his family, as will be discussed later.

Gross

MTC is typically a firm rounded tumor which is often located within the middle or uppor lobes of the thyroid gland [220, 254, 270, 271, 537], a position corresponding to the greatest concentration of C cells in the normal human thyroid [346, 547]. However, the distribution may be more variable and a case involving the isthmus, an area relatively free of C cells, has been described [333]. Microscopic foci of tumors are found at sites remote from the main tumor mass [333]. This multifocal pattern of neoplasia is especially common in familial MTC [298, 352]. In familial cases the tumor is almost always bilateral, but it is also bilateral in the majority of apparently sporadic cases [127, 254, 334]. This greater tendency to bilaterality is the only pathological or histological difference between familial and sporadic MTC. Depending upon its stage, the tumor can range in size from a few millimeters up to 13 cm in diameter [550]. On cut surface, the tumor is grey-white in appearance and gritty in texture; the latter characteristic is due to commonly occurring calcification [537]. The tumor is either fully or partially encapsulated, or well demarcated, but it may be invading adjacent parenchyma and blood vessels [17, 242, 259, 270, 271, 293, 333, 537].

Pathology

Histology

The histological elements of the tumor can have a highly variable appearance which may change during the course of the disease [200]. Although some histological features have been related to prognosis, and these are discussed subsequently, there is no consistent relation between histological appearance and biological behavior [63, 254, 333]. There are four histological features which can contribute to the appearance of the tumor: (a) the cellular elements; (b) the stromal elements; (c) the presence of amyloid; and (d) calcification (fig. 1).

The cells are usually polyhedral or polygonal in shape but occasionally round. They have a finely granular eosinophilic cytoplasm. Electron-microscopic (EM) studies reveal the presence of secretory granules, presumably calcitonin, in the cytoplasm [89, 238]. The nucleus is usually central in location and the nucleolus not prominent. Mitotic figures are usually rare but binucleate and multinucleate cells are common. A less common cytological feature is the presence of spindle-shaped cells, but these can predominate in some cases [333]. *Williams* et al. [537] reported that spindle cells along with increased mitoses were unfavorable prognostic elements. *Ibanez* et al. [270, 271] felt that polyhedral cells carried a better prognosis. Cells intermediate to polyhedral and spindle cells have also been described.

The cellular elements can be arranged to form a variety of patterns. These patterns have been variably described as sheets, ribbons, palisades or streaming, organoid or alveolar, trabecular, angiomatous, whorled, and pseudorosette [254, 270, 271, 333, 537]. There can even be follicular and papillary structures. The latter two formations are very rare and their presence is more consistent with the diagnosis of follicular or papillary carcinoma of the thyroid. *Ibanez* et al. [270, 271] demonstrated the presence of follicular and papillary elements in 3 of 57 cases of MTC, and shrinkage of MTC can give a pseudopapillary appearance. A giant-cell variety of the tumor has also been described [286]. It is extremely rare to see MTC occurring simultaneously along with other forms of thyroid cancer, but such mixed tumors have been reported.

The arrangement of cells can be influenced by the distribution of stromal elements. In some tumors, the stromal elements are scanty, in others intermediate, and in still others, predominant. *Keiser* et al. [298] suggested that the development of stromal elements parallels the age of the tumor. They observed that cellular elements were predominant in younger patients

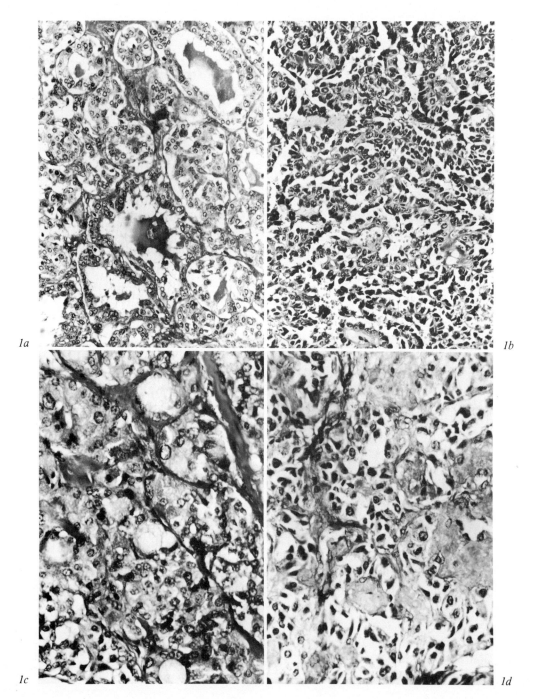

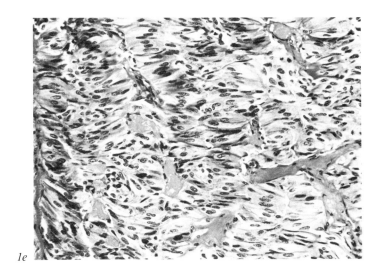

1e

with the tumor, whereas fibrosis was more common in older patients with the tumor, all of whom had cervical-node involvement.

It is interesting to note that the histological appearance of MTC can resemble that of other tumors of neural-crest origin which occur in association with MTC and which may also produce calcitonin and/or the other bioactive substances produced by MTC. These tumors include oat-cell carcinoma of the lung, carcinoid tumors, islet-cell tumors, angiomata, pheochromocytoma, neurolemmomas, neuromas, paragangliomas, neurogenic sarcomas, chemodectomas, melanomas, and carotid and aortic-body tumors [270, 271, 333, 359, 537]. These observations provide a further link for the tumors of neural-crest origin.

Fig. 1.
- a Medullary thyroid carcinoma (MTC) with follicular structures.
- b MTC with papillary structures.
- c MTC with large vacuolated tumor cells.
- d MTC with clusters of loosened cells with frayed cytoplasm and intermingled with small clumps of amyloid.
- e Medullary carcinoma composed of palisading spindle cells. Small amounts of amyloid are present in connective tissue septa [from ref. 333].

Amyloid

A more constant feature of MTC (and its metastases) than the histocytological appearance of the tumor is the presence of amyloid. Amyloid occurs with such regularity in MTC that its absence, if a large-enough area is surveyed, must be considered a strong point against the diagnosis. However, amyloid may not always be present [385] and its importance in the diagnosis of MTC has been reduced and may be replaced by immunochemical staining procedures which specifically demonstrate abnormal C cells [270, 546]. Furthermore, amyloid may be present in the thyroid tumors which are not MTCs [511]. Most of the evidence suggests that amyloid is produced by the C cells of the tumor. The amyloid is present both intracellularly and extracellularly. Intracellularly, it occurs in association with the cytoplasmic granules of the cells [359, 499]. Even MTC cells grown in tissue culture may produce amyloid [7]. However, the amount of amyloid varies among tumors. It can be unevenly distributed within a tumor and large areas may be devoid of amyloid. Amyloid may be present among the stromal and cellular elements of MTC [254]. The deposits of amyloid can be small or large, and occasionally, massive depositions of amyloid can be seen. *Ibanez* et al. [270, 271] suggested that tumors with more amyloid had a better prognosis. *Hill* et al. [254] observed that the polyhedral cells are more likely to have amyloid than spindle cells. Some amyloid deposits have a laminated appearance and others exhibit birefringence with polarized light [333]. The presence of multinucleated giant cells can suggest a foreign-body reaction [254, 270, 271]. *Tubiana* et al. [510] reported the existence of an MTC variant which was amyloid-free and ran a more malignant clinical course.

The amyloid seen in MTC has staining, histochemical, and ultrastructural characteristics similar to those of primary, secondary, and experimental amyloidosis [7]. There is, however, some recent evidence which indicates differences between the immune amyloids and the amyloid of MTC [226, 482, 532]. These data demonstrate biochemical, immunochemical, and structural differences between these two forms of amyloid. The structural data, partial amino acid characterization, have been further interpreted to indicate that there are structural similarities between the amyloid of MTC and human calcitonin [532]. If this preliminary evidence is confirmed, it would support the hypothesis of *Pearse* et al. [410] that the amyloid seen in a variety of endocrine tumors, such as insulinomas, carcinoid tumors, parathyroid adenomas, and MTC, is structurally related to the secretory product of the particular gland [301]. For example, they suggested that the amyloid of insulinomas may be related to the carboxy-peptide chain of insulin. This phenomenon may have a counterpart in non-neoplastic endocrine cells since microdeposits of amyloid occur in normal pituitary glands, islet cells, adrenal and thyroid glands [428]. The report of amino-acid similarities between MTC amyloid and islet-cell amyloid further suggests that there might be structural similarities among the endocrine amyloids [532].

Although the majority of evidence thus indicates a C-cell origin for the amyloid of MTC, this view has been challenged by the recent work of *Thiliveris* et al. [506]. Their EM and histochemical studies demonstrated the presence of amyloid in association with fibroblasts of a thyroid tumor, tongue neuroma, and eyelid neuroma in a patient with MTC. Their interpretation was that the fibroblasts rather than C cells deposited the amyloid seen in this patient's tumors. In a few cases of MTC, amyloid has also been observed in nonthyroidal sites. However, this has been such a rare occurrence in contrast to the common finding of amyloid in MTC that it is likely that the nonthyroidal amyloid in these patients was related to factors other than the thyroid tumor. Conversely, *Valenta* et al. [511] observed the presence of amyloid in a tumor resembling MTC but subsequently shown not to be MTC.

Tumor Calcification

Calcification is commonly found in MTC, with an incidence of 50% reported in one series [333]. *Williams* [539] associated a small amount of calcification with poor prognosis. Calcium deposits are commonly found in the fibrous stroma or the amyloid deposits of the tumor. Calcification lacks concentric lamination of psammoma bodies [270, 271]. The calcifications are also more dense and irregular than the faint, homogeneous psammoma bodies which occur in other thyroid cancers and the cystic calcification found in goiters. Local and distal metastases may also calcify, although lung and mediastinal metastases do not commonly calcify [298, 333]. True ossification has also been reported in a few MTCs; the bone foci consisted of bone trabeculae enclosing bone marrow with blood-forming tissue [333].

In addition to their pathological interest, the depositions of calcium within the MTC may also be of clinical value. In a significant percentage of tumors, the calcifications are of sufficient magnitude to be detected by X ray [521]. In one series, dense calcifications were present in 7 of 19 patients with MTC [412] and in 8 of 20 patients in another series [298]. Therefore, the appropriate soft-tissue films of the neck can be a diagnostic adjunct for the patient with MTC.

C-Cell Hyperplasia (CCH)

In *Williams'* [536] description of the histogenesis of MTC, he observed that foci of hyperplastic C cells occurred in the rat form of this tumor and that it was difficult to distinguish between such hyperplastic areas and true tumors. *Ljungberg* [331, 332] described increased C-cell populations remote from the tumor in MTC. In this setting, *Wolfe* et al. [546] described CCH as a distinct pathological entity. They were studying 3 patients at risk for MTC because of their family history. These patients had small but progressive increases in plasma calcitonin during calcium infusion and consequently underwent thyroidectomy. The extirpated thyroid glands did not reveal the presence of MTC. However, there were clusters of hyperplastic parafollicular cells which were calcitonin-positive by immunohistological studies. The presence of increased calcitonin in these cells was confirmed by bioassay and immunoassay. The distribution of these hyperplastic parafollicular cells was in the area of the thyroid gland where C cells are usually most prominent, the upper and middle portions of the lateral thyroid lobes. These observations suggested that, at least in familial cases of MTC, the frank malignancy is preceded by a progressive hyperplasia of C cells. These cells exhibited no nuclear atypia or invasive tendencies [244].

Although cellular hyperplasia has been suggested to be a precursor for many forms of malignancy, this progression has seldom been so well documented as it is for MTC. The variable behavior of MTC is reflected in this predecessor of the tumor in that the CCH can become manifest as early as 2 years [118, 127, 298] or as late as 23 years of age [546]. Furthermore, *Normann* [386] has reported the diagnosis of CCH being made beyond the third decade. The finding of these early stages of MTC, in addition to its fundamental importance for cancer pathogenesis, is of considerable clinical significance. The early stages of MTC, when the neoplastic process is confined to the thyroid gland, are the most amenable to surgery; CCH and even more subtle histological changes are below the threshold of clinical detection, but may be identifiable with the calcitonin assay. Such early identification offers the best hope for effective therapy and cure [120].

C-Cell Adenoma

An adenoma of thyroidal C cells has only rarely been reported [69, 364]. In 1 case identified in a series of 12 nontoxic thyroid adenomata, the cells were identified as C cells on the basis of their staining characteristics only. The adenoma did not concentrate radioactive iodine in vivo and an extract of it produced hypocalcemia in the rat. In another patient, diarrhea disappeared after the removal of a C-cell adenoma [364]. These results must be considered as only preliminary evidence for the existence of adenoma of C cells. Definitive evidence will be provided by specific immunohistochemical studies which demonstrate the presence of calcitonin in tumor and perhaps in peripheral blood.

Natural History

MTC is generally regarded to be intermediate to the aggressive behavior of anaplastic thyroid carcinoma and the more prolonged course of papillary and follicular carcinoma. In three series the 5-year survival ranged from 48 to 55% [254, 270, 351]. Of more clinical importance than the survival data for large groups of patients is the variability in survival among individual patients. Survival can vary from months to 30 years after diagnosis [189, 402, 420, 446]. *Mandelstam* et al. [339] reported a 4.5-year-old patient with MTC and metastases, but with a palpably normal thyroid. Patients under 2 years of age with metastatic disease have been reported, as has a patient of 52 years of age with only localized disease [53, 195, 490]. CCH has been diagnosed as early as 2 years [385] and as late as 45 years of age [385, 479]. Therefore, the tumor can be rapidly aggressive, leading to death within months after diagnosis, or it can be indolent and compatible with decades of life.

Metastases to cervical and mediastinal lymph nodes are very common, being present at the time of diagnosis in up to 50% of patients [351]; this can occur even if the primary is less than 1.0 mm in size. Distal metastases favor the lung where they can present as diffuse nodular lesions with a fibrotic pattern [521]. Metastases to liver, which can occur in up to 25% of patients, distinguish the behavior of this thyroid tumor from others [302]. In one series, 18% of patients had bone metastases [254]. Although 30% of these metastases were predominantly blastic, all the metastases had lytic components. The bone-resorbing characteristics of the tumor were apparently great enough to overcome the potential for inhibition of bone resorption of the increased (local and systemic) calcitonin.

There is no consensus regarding the relationship between the histological appearance of the tumor and its clinical behavior. *Williams* [539] associated a poor prognosis with (a) spindle cells, (b) areas of necrosis, (c) little calcification, and (d) frequent mitoses. *Hill* et al. [254] and *Ibanez* et al. [270] could not confirm these observations. In general, the tumor is considered to have an intermediate grade of malignancy compared with other thyroid tumors, but survival is very variable, ranging from months to decades. There is some evidence to suggest that MTC occurring within a given family has a consistent biological behavior and under these circumstances some tumors have a low malignant potential [256].

Calcitonin

Biochemistry

Both the amino-acid composition and sequence of seven types of calcitonins are known [487]. While the first crude extracts came from the thyroids of rats [257, 258], porcine thyroids provided sufficient hormone for biochemical studies of composition [420] and sequence [378]. Porcine calcitonin was revealed to have a 32-amino-acid sequence which was subsequently synthesized [239, 432]. Human calcitonin was isolated [431] from MTC, and shortly thereafter it was sequenced [377] and synthesized [463]. Presently, calcitonin structures of bovine [90], ovine [452], salmon [300, 301, 380], rat [99, 472], and eel [176, 381] have been identified.

While there is a certain amount of commonality in the features of naturally occurring calcitonins, there is also considerable variability. Common features include the 1,7 amino-terminal disulfide bridge and the carboxyl-terminal prolinamide residue. 7 of the 9 amino-terminal residues are identical in all calcitonins; however, only 2 sequence positions (residue 28, glycine, and residue 32, prolinamide) are completely conserved. Generally, the 10–27 region is the most homologous among porcine, bovine, and ovine calcitonins; between human and rat calcitonins, and between salmon and eel calcitonins, respectively. Human calcitonin is more similar to salmon calcitonin (17 of 32 residues in common) than to porcine (12 of 32 residues in common). Based on cross-reactivity, it appears that there is considerable homology between salmon and chicken calcitonins [153, 154].

Studies of calcitonin extracted from the UB glands of a variety of salmon species reveal considerable microheterogeneity. Each species contains two forms of calcitonin, designated salmon calcitonin I and either II or III. Salmon calcitonin II differs from calcitonin I in positions 15, 22, 29, and 31. Salmon III is identical to calcitonin II in these positions but differs from I at residue 8, where valine is replaced by methionine [301].

The apparent variability in the middle region of calcitonin can largely be explained in terms of single base exchange, presumably through mutation. There is considerable similarity when comparison is based on the chemical properties of the amino-acid side chains. For example, acidic residues (aspartic or glutamic acids) are consistently found only at position 15, although an acidic residue is also found at position 30 in porcine, bovine, and ovine calcitonins. Similarly, basic residues are confined to relatively few positions. Where basic residues have been substituted, asparagine and glutamine are the most common replacements. Since amides are weakly basic, this exchange is considered conservative, and is quite common in other groups of related proteins and peptides. Leucine, phenylalanine, or tyrosine (hydrophobic residues) distributed rather regularly along the peptide chain are in positions 4, 9, 12, 16, 19, and 22.

There is conservation at a number of sites. All calcitonins contain at least one acidic residue at positions 15 and/or 30, and there is moderate conservation in the distribution of basic residues as well. An amide or a basic residue (arginine, lysine, or histidine) is consistently found in positions 14, 17, 18, and 20, and all calcitonins contain at least two basic residues. While methionine is generally found in positions 8 and 25, it is apparently not required for biological activity since it is lacking in salmon and eel calcitonins. Tryptophan is found in porcine, bovine and ovine calcitonins at position 13, but not in salmon, human, rat, and eel. Tyrosine is absent in salmon II and III, but is present at three sites in ovine calcitonin. When tyrosine is present, it is found at positions 12, 19, and/or 22. When tyrosine is absent, it is replaced by a hydrophobic residue.

There appears to be a relationship between hormone potency and the degree of basicity. The UB calcitonins, which are more basic, are also more potent. Similarly, basic substitutions in human calcitonin enhance potency [338]. Conversely, potency can be diminished by deletion of the C terminus and opening the disulfide ring. Potency is unaffected by N-terminus modifications [433, 464]. There is also evidence suggesting that potency correlates well with stability to plasma enzymes. Salmon calcitonin is both more stable and more potent than the mammalian calcitonins. A biologically active calcitonin fragment has not been identified, suggesting that the whole peptide is involved in the biological activity, with tertiary structure and chemical groupings at the termini being particularly critical to effective receptor binding [419].

Calcitonin Secretion by Normal C-Cells

Calcium and Related Minerals
The secretion of calcitonin is directly related to blood-calcium concentrations [159, 400]. When blood calcium rises acutely, there is a proportional increase in plasma calcitonin. An acute decrease in blood calcium produces a corresponding decrease in plasma calcitonin. This is true both in humans and experimental animals [174, 181]. There is evidence that women have decreased calcitonin response to calcium stimulation compared with men [246, 401]. Thus, the changes in blood calcium and plasma calcitonin seen with acute challenge are well established. However, the effects of chronic hypercalcemia and hypocalcemia are not fully defined, and contradictory results have been reported. The variable responses observed in patients with hypercalcemia may reflect a varying capacity of C cells to respond to chronic stimulation. Animal studies indicate that C cells can become exhausted during chronic exposure to elevated blood calcium [206]; however, these findings have many limitations in their applicability to humans.

When calcitonin is elevated in patients with hypercalcemia and malignancy, it has been difficult to attribute the elevation to hypercalcemia. Other possible causes include ectopic calcitonin production by the tumor or other sites and impaired renal function [135, 469]. Normal, elevated, and decreased levels of calcitonin have been reported in hyperparathyroidism and many studies have failed to demonstrate any consistent abnormality [5, 253, 314, 374, 466, 501]. However, in some hyperparathyroid patients, histological evidence of CCH suggests increased calcitonin secretion [310, 329, 334, 500]. Furthermore, recent studies have demonstrated increased plasma-calcitonin secretion in patients, primarily males, with primary hyperparathyroidism [402], whereas calcitonin reserves in females with primary hyperparathyroidism were decreased, as they are in normal females [401, 402].

Chronic hypocalcemia does not appear to have a consistent effect on calcitonin secretion, although some studies suggest that chronic hypocalcemia can suppress secretion [165]. In these studies of hypocalcemic patients, both calcium and pentagastrin infusion produced a greater increase in plasma calcitonin than in normal subjects. Chronic hypocalcemia may thus result in increased storage of calcitonin in C cells, which is released by secretagogues. While increased thyroidal calcitonin has been described in chronic hypocalcemic states [10], additional studies of secretion are needed to confirm this proposed mechanism.

Calcitonin secretion appears to be acutely stimulated by minerals related to calcium [137]. Both magnesium and strontium have demonstrated this action in animal studies, although less strongly than calcium [109, 414]. Magnesium has been shown to stimulate calcitonin secretion in normal human subjects as well [174], but in some patients with MTC an opposite effect has been observed [14].

Gastrointestinal Factors

Intravenous administration of pentagastrin (or of the closely related cholecystokinin, tetragastrin, and caerulin) and glucagon produces a significant increase in plasma calcitonin [36, 43, 163, 401]. However, in these studies the dose of hormones was sufficiently large to increase blood concentration to levels higher than those seen even in the postprandial state. Thus, the physiological significance of the regulation by these hormones of calcitonin secretion is unclear [112]. Moreover, an increase in plasma calcitonin following eating and/or an oral calcium load has been demonstrated in only a few clinical and experimental studies [139, 390]. Thus, the secretory relationship between the gastrointestinal tract and C cells needs further exploration to determine its physiological significance [108, 110].

Neuroendocrine Factors

Adrenergic agents can both stimulate and suppress the secretion of calcitonin [106]. The ability of neuroendocrine factors to modulate plasma-calcitonin levels has been observed in clinical studies in normal subjects and in patients with primary and secondary abnormalities of calcitonin secretion. Calcitonin secretion can be stimulated in humans by the beta-agonist isoproterenol, and the alpha-antagonist, phentolamine. The alpha-antagonist, methoxamine, has also been reported to stimulate calcitonin in humans [47, 48, 358, 518]. However, these studies do not preclude a possible secretory effect due to changes in blood flow to the thyroid, small changes in ionized calcium, or alterations in hormone metabolism. The beta-antagonist, propranolol, has been reported to decrease calcitonin levels. Oral *L*-dopa has been reported to depress plasma-calcitonin levels in normal humans and in patients with MTC [61]. While these results suggest that dopa uptake and decarboxylation to dopamine may play an inhibitory role in calcitonin secretion, recent studies of bromocriptine failed to consistently elicit a similar response and *L*-dopa caused suppression only in a minority of normal and MTC patients [486].

The effect on calcitonin of somatostatin remains to be clarified. High doses of somatostatin in animals suppress calcitonin [240], but lower doses in human studies have not demonstrated a similar effect [358]. Since somatostatin-containing cells have been demonstrated in the thyroid by immunohistological techniques [261], the relationship between the two hormones may be significant. Further study is needed to establish both the nature and the significance of the relationship between neuroendocrine factors and calcitonin secretion.

Age and Sex

While age and sex seem to affect plasma-calcitonin levels, their specific role appears to be species-dependent. In salmon, plasma calcitonin is higher in females [524], but in bovine species the opposite is true. Also, in bovines, calcitonin is decreased in older animals and increased in those on high-calcium diets, thus suggesting a role for age and dietary factors [164]. In rats, females have higher plasma-calcitonin levels than their male counterparts, and hormone concentration increases with age [441]. Early evidence suggests an influence of both sex and age on calcitonin secretion in humans. Specifically, higher-than-adult plasma levels have been reported in pregnant females, children, and newborn infants [450]. Menstrual effects on plasma calcitonin have also been observed, but not consistently [42, 416]. Levels also appear to decrease with age in humans, in contradistinction to rats [157, 185, 447], and it is well established that calcitonin secretion is decreased in men compared to women (fig. 2) [246, 401, 402, 459]. Thus, a variety of species-specific factors appear to modulate the effect of age and sex on calcitonin secretion.

Other Factors

There appears to be a relationship between calcitonin secretion and iodine metabolism in experimental animals. This is significant because C cells have not been classically considered to play a role in iodine metabolism [259]. However, in iodine-deficient rats, CCH is observed and plasma calcitonin is elevated; the effect can be reversed through replacement of iodine and perhaps by thyroxine [130]. In some cases, the observed effect may be due to the development of hypercalcemia [130]. By contrast, athyreotic cretins may show decreased calcitonin secretion [111]. While the effect is difficult to reproduce in humans, high doses of prostaglandins stimulate plasma calcitonin in rats [358, 439]. There is some evidence which suggests an autoregulatory effect on its secretion by calcitonin [67]. The secretion of calcitonin, as is true for many peptide hormones, is mediated by adenylate cyclase [67, 247].

Secretion by MTC

MTC is a neoplastic disorder of the C cells of the thyroid gland. Thus, the tumor produces abnormally high amounts of calcitonin, exceeding normal thyroid content by orders of magnitude. As a result, patients with this tumor have elevated concentrations of calcitonin in their peripheral blood and urine. In many patients, basal concentrations of the hormone are sufficiently elevated to be diagnostic of the tumor's presence. Therefore, the radioimmunoassay for calcitonin can be used to diagnose the presence of MTC with an exceptional degree of accuracy and specificity when applied to

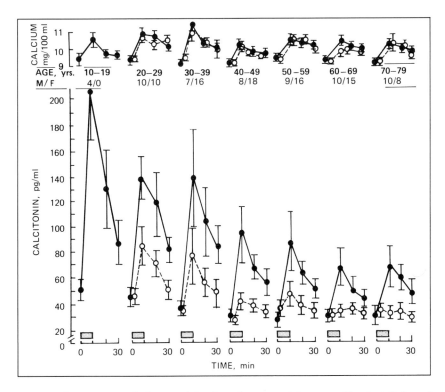

Fig. 2. Responses of plasma calcitonin and serum calcium to a 10-min infusion of calcium in 58 normal males (●) and 83 normal females (o). Stippled bars denote infusion of calcium (as the chloride salt) at a dosage of 3 mg/kg body weight. Values represent mean ± SE for each determination. Indicated at the top are the numbers of male and female subjects in each decade. Plasma calcitonin is greater in males than in females ($p < 0.05$–0.001) and there is a progressive decrease with age in both sexes. To convert calcium values to millimoles per liter, multiply by 0.25 (from [185]).

measurements in plasma samples. In a small but increasing percentage of patients, however, basal hormone levels are indistinguishable from normal levels [166]. Many of these cases represent early stages of C cell neoplasia or hyperplasia, and are most amenable to surgical cure. To identify these patients with early disease, provocative tests for calcitonin secretion have been developed which have led to the identification of MTC in patients whose diagnoses could have been missed if only basal calcitonin determinations had been considered.

Provocative Testing
Glucagon
Glucagon can stimulate calcitonin secretion, but it is not generally used as diagnostic agent. Glucagon is not a reliable calcitonin secretagogue because its effect is variable [163]. Glucagon can release catecholamines, and since patients with MTC may have associated pheochromocytomas an adrenergic crisis can be precipitated by its administration [174].

Calcium
Intravenous infusion of calcium has been the most widely used diagnostic agent for MTC. In early studies, calcium was infused at doses ranging from 3 to 5 mg/kg/h for periods varying from 2 to 4 h. The increase in calcium produced by such infusions, usually several milligrams per deciliter, consistently produced an abnormal increase in plasma calcitonin in MTC patients [501]. The abnormal increase in plasma calcitonin occurred even in tumor patients who had basal concentrations of the hormone that were indistinguishable from normal [166]. Thus, the measurement of calcitonin after calcium infusion served to clearly identify MTC patients [166, 345].

Prolonged calcium infusions have disadvantages. The procedure is fairly long and the dose of calcium often produces untoward effects such as hypertension, nausea, and vomiting. This has necessitated hospitalization in research wards. More recently, shorter infusions have been developed [399, 444]. In these procedures, the increase in plasma calcium is approximately 1 mg/dl, yet calcitonin release is reliably stimulated. The procedures can be completed in a few minutes and are relatively well tolerated [166, 399].

Pentagastrin
Pentagastrin is another widely used provocative agent for calcitonin secretion in patients with MTC [250]. When administered by rapid intravenous injection, pentagastrin produces a prompt increase in plasma calci-

tonin [136]. The response probably is a function of the dose and speed of injection, inasmuch as calcium, when administered in a similar fashion, produces a response similar to that observed with pentagastrin, whereas infusion of pentagastrin over several minutes produces a response similar to that produced by a calcium infusion [174, 444]. The rapidity of pentagastrin infusion also has drawbacks, as it leads to an unpleasant sensation, commonly described as burning or flushing. Moreover, since the use of pentagastrin as a diagnostic tool for suspected MTC is not currently approved by the Food and Drug Administration, an institutionally approved protocol may be required for its administration. The use of tetragastrin as an agent for calcitonin secretion has been determined to be less effective than pentagastrin [5].

An interesting modification of the pentagastrin test has been described [528]. Patients were given the peptide while plasma samples were taken via an indwelling catheter in their inferior thyroid veins. Calcitonin was measured in the plasma samples, demonstrating a dramatic and diagnostic increase in thyroidal vein calcitonin. In some patients this increase could be detected even when peripheral calcitonin did not increase. However, such procedures cannot yet be considered routine, as the concentration of thyroidal venous calcitonin is not well established and can be influenced by a small change in the position of the indwelling catheter. Also, more sensitive radioimmunoassays have better defined normal and abnormal ranges of basal as well as stimulated calcitonin concentrations in peripheral human plasma [401]. Suspected MTC patients that had required provocative testing may now be clearly identified by basal calcitonin measurements with assays of improved sensitivity.

A more practical use of selective venous catheterization is in the evaluation of the location and extent of MTC or an ectopic calcitonin-producing tumor [224, 528]. The procedure can help localize a recurrent MTC and lead to more effective treatment [224]. However, high circulating-hormone levels may obscure the resulting calcitonin gradient, even when this difficult procedure is technically successful.

Pentagastrin versus Calcium

The relative merits of pentagastrin and calcium have been discussed [250, 528]. Most tumors respond to either agent and there are few false-negative results. However, some tumors respond to calcium but not to pentagastrin, and vice versa [221]. Thus, both procedures should be considered before MTC is ruled out [174]. So far, combined calcium-pentagastrin infusions have not yielded consistent results [345, 528]. However, preliminary evidence suggests that calcium infusions may be more valuable for the diagnosis of early forms of the tumor [345]. The key to both of these procedures is the sensitivity and specificity of calcitonin assay [327].

Whiskey

Scotch whiskey has been identified as a provocative agent for calcitonin [201]. In patients with MTC, imbibition of 50 ml elicits an increase in calcitonin comparable to that seen following some calcium infusions; the effect is generally seen within 15 min. The high incidence of false-negatives suggests, however, that more experience with this agent is necessary. Despite its ease of administration, the possibility of diarrhea and flushing suggests additional untoward effects in the MTC patient [132].

Magnesium

There is evidence to suggest that magnesium affects calcitonin secretion of normal individuals and experimental animals differently than it does patients with MTC [14]. In the former group, magnesium infusion elevates plasma calcitonin, while in 3 patients with MTC, magnesium suppressed calcitonin secretion. This may suggest a difference between normal and malignant C cells in their response to magnesium.

Venous Catheterization Procedures

Selective venous catheterization can be used in conjunction with calcitonin assay to establish the location as well as the presence of MTC. This is accomplished through the analysis of the hormone-concentration gradient in a specific vein in order to localize the tumor to the site draining that vein. Accurate catheter placement and confirmation with venography are critical to the success of the technique. However, even with care, the procedure has limited usefulness in primary diagnosis because of the high incidence of bilaterality in MTC. To use the procedure to evaluate tumor persistence and recurrence postsurgery, catheterization must be of an area in which the venous pattern has not been distorted by the surgery. This necessitates arteriography to establish blood-flow patterns, a procedure which adds considerable risk, time and expense to the catheterization. Therefore, the greatest potential benefit for catheterization is in the location of metastatic disease, the knowledge of which can greatly influence selection of therapy [26, 224].

Serial Calcitonin Measurements in the Evaluation of Therapy

Monitoring of plasma calcitonin is an effective means of evaluating therapy for calcitonin-producing tumors [173, 236]. In postsurgical patients, levels fall sharply, but may reach a low point only 1–6 months following excision [489]. In patients treated radiologically, high calcitonin levels persist for an average of 3 years. There is often a marked increase prior to metastasis and death. Highest levels are associated with patients with hepatic involvement, but, in general, calcitonin levels correspond to the extent of disease, and recurrence can be monitored through both basal and provocative testing [173, 221, 355, 467, 489].

Immunochemical Heterogeneity

There are multiple immunochemical forms of calcitonin in tumors and in plasma [169, 225, 475]. When the plasma or tumor of a patient with MTC is immunoassayed after gel-filtration chromatography, several peaks of immunoreactive calcitonin are seen. Their number is a function of the size of the column, the nature of the matrix gel, and the elution conditions. Calcitonin auto- or homoaggregation (e.g., dimerization or polymerization) or heteroaggregation (e.g., with other proteins) also affect the elution profile. Thus, only some of the peaks actually reflect the biosynthesis, secretion, and metabolism of calcitonin. Since biosynthetic precursors for calcitonin have been described [438], a preprocalcitonin or procalcitonin may also be represented in plasma [13].

The immunochemical profile of calcitonin may vary among certain disease states. This variation has been demonstrated for calcitonin-producing tumors and for renal disease [174, 322]. The various species of plasma calcitonin seen in renal disease and malignancy are produced in part by abnormalities in hormone metabolism and secretion. Therefore, the calcitonin peak representing the monomer seen in MTC plasma is nearly absent in renal disease where most of the calcitonin elutes at or near the void volume [322]. This also appears to be true in some instances of ectopic calcitonin production where the hormone may also elute at or near the position of the biosynthetic precursor(s) of calcitonin [13, 438]. In other malignancies the calcitonin monomer may predominate. Therefore, it may become possible to use the immunochemical elution pattern of plasma calcitonin for the differential diagnosis of hypercalcitoninemic states. Thus, both specificity and sensitivity must be considered in assessing the diagnostic value of a calcitonin assay system.

Antibodies of a given specificity for the calcitonin molecule will preferentially react with hormone species which express that specificity. For example, if calcitonin is metabolized to a fragment that contains a carboxy-terminal peptide, an antibody with specificity for the carboxy-terminal region of calcitonin will recognize and have a higher affinity for that fragment. Immunochemical heterogeneity is thus a function of the hormone species being measured and a function of the analytical procedure.

Assessment of the immunochemical heterogeneity of plasma calcitonin may also provide information about C cell function because it will indicate whether excess production occurs at the prehormone, hormone, or metabolic product level. However, the most important clinical use of calcitonin heterogeneity is in the screening of patients in the early diagnosis of MTC or other calcitonin-producing tumors. Some assay procedures, because of their immunochemical specificity or sensitivity, may more effectively identify slightly increased basal concentrations of calcitonin which occur early in the course of neoplasia. Furthermore, different provocative agents may stimulate the secretion (or release) of different species of calcitonin [468, 484]. Thus, the optimal diagnostic procedure is one which has a sensitive assay which is specific for measuring the species secreted in response to a provocative test.

Metabolism

The metabolism of calcitonin is a complex process which involves many organ systems. Evidence has been reported for degradation of the hormone by kidney, liver, bone, and even the thyroid gland. Unfortunately, many of the published data regarding calcitonin metabolism have questionable physiological significance. Most of the studies have been done with either pharmacological amounts or radioiodinated species of calcitonin [22, 129]. Furthermore, many studies were performed with nonhomologous species of calcitonin [129, 474] and many studies have assessed metabolism by measuring the immunological activity of calcitonin rather than biological activity [22, 474]. For these reasons, only tentative conclusions can be drawn regarding the metabolism of calcitonin. Like many other peptide hormones, calcitonin disappears from plasma in a multiexponential manner which includes an early half-life measured in minutes [22, 129]. The rapid component(s) may represent mixing of calcitonin within its volume of distribution and the slower component(s) may represent the rate of true decay. In most studies, the kidney seems to be the most important organ of clearance for calcitonin [22, 129]. Inactivation of the hormone seems more important than renal excretion since relatively little calcitonin can be detected in urine [129]. Liver is also an important organ in calcitonin clearance and in some studies it is dominant [474]. Bone, thyroid, and plasma degradation also contribute to the disappearance of calcitonin from the circulation [60, 129, 474].

Clinical Uses of Calcitonin

The therapeutic applications of calcitonin utilize its major biologic actions of inhibiting bone resorption and lowering blood calcium. These two effects of calcitonin make it useful in the treatment of patients with diseases characterized by increased bone resorption. Calcitonin has thus been used in the treatment of Paget's disease, osteoporosis, renal osteodystrophy, and osteogenesis imperfecta [30, 38, 121, 183, 365, 375]. In Paget's disease calcitonin treatment is quite effective in reversing the abnormal biochemical, clinical, and radiological manifestations of the disorder. In hypercalcemic states, calcitonin has been used in the treatment of hyperparathyroidism and hypercalcemia associated with malignancy. Calcitonin is very effective in the latter, and its hypocalcemic effect can be enhanced by the use of steroids and phosphate [182]. Three forms of calcitonin have been used: porcine, salmon, and human. Porcine calcitonin has been most widely used in continental Europe, human calcitonin in England, and salmon calcitonin in the United States. Salmon calcitonin is the most potent form of the hormone. However, it is a foreign protein and produces antibodies in humans which may inhibit the therapeutic effects of the hormone [473]. The availability of human calcitonin should widen the therapeutic applications of calcitonin [549].

Differential Diagnosis of Hypercalcitoninemia

Elevated levels of plasma calcitonin were considered to be diagnostic of MTC when the features of this tumor were first being elucidated. However, it has become apparent that other physiological and pathophysiological states may be associated with elevated plasma calcitonin. The growing appreciation of the complexity of calcitonin biosynthesis, secretion, and metabolism has increased the number of conditions which must be considered in evaluating the patient with abnormal plasma calcitonin. In some of these conditions, such as pancreatitis and pregnancy, there still remains conflicting evidence regarding the occurrence of hypercalcitoninemia. If it does occur, it is a transient phenomenon and its clinical setting is not likely to cause a diagnostic problem. In other conditions, such as ectopic calcitonin production by a nonthyroidal tumor, the differential diagnosis of hypercalcitoninemia may be more difficult.

Hypercalcitoninemia Associated with Malignancy

This diagnostic category can be divided into 2 groups of patients – those with ectopic calcitonin production by a nonthyroidal tumor and those in which the malignancy is associated with abnormal calcitonin production by the thyroid.

Ectopic Calcitonin Production
It is now clearly established that cancers other than MTC can produce abnormal concentrations of calcitonin [255]. This observation was first made in carcinoid tumors which are embryologically related to MTC and which can, in fact, occur in association with MTC [363]. Other tumors associated with ectopic calcitonin production such as oat-cell carcinoma of

the lung and islet-cell carcinoma of the pancreas, are also likely to represent tumors of cells from neural crest origin which are, therefore, embryologically related to MTC [27, 173]. Therefore, many of the tumors associated with ectopic calcitonin production probably derive this potential from their common embryological origin with MTC [64, 66, 231, 292, 507].

Of these groups of tumors, the most clinically significant is oat-cell carcinoma of the lung. Although not the most common carcinoma of the lung, it is being reported with an increased frequency and it is the subject of much interest in the chemotherapy of cancer. When this information is combined with the common propensity for oat-cell carcinoma of the lung to secrete calcitonin along with a variety of other biologically active substances, the potential importance of the calcitonin assay in the diagnosis and management of patients with this tumor becomes obvious [470, 522].

Eutopic (Thyroidal) Calcitonin Production

Although ectopic calcitonin production by the tumor can explain the hypercalcitoninemia seen in some patients with cancer, alternative explanations are necessary for many of the reported instances. In contrast to tumors such as oat-cell carcinoma of the lung, islet-cell carcinoma of the pancreas, and some carcinoid tumors, it is not likely that the hypothesis of neural-crest origin for the cell type of the tumor can be used to explain the wide variety of tumors associated with hypercalcitoninemia, the most notable of which is carcinoma of the breast [255]. Several alternative explanations have been suggested for this association. It has been demonstrated that in some such cases, the thyroid gland, not the tumor, is the source of the excessive calcitonin [467]. Another explanation is that the tumor metastasizes to bone, resorbs bone, and produces an autoinfusion of calcium which can stimulate thyroidal calcitonin secretion. This hypothesis is supported by the presence of hypercalcemia in some patients with hypercalcitoninemia and metastatic disease. However, hypercalcemia is not an invariable finding in the patients with hypercalcitoninemia. Another hypothesis is that the tumor produces a calcitonin secretagogue. For example, there is experimental evidence to suggest that prostaglandins may stimulate calcitonin secretion, and it is well known that a variety of tumors produce excess prostaglandins [439, 540]. However, this mechanism has not yet been demonstrated in patients with hypercalcitoninemia associated with malignancy. Gastrin production by the tumor is also a possible mechanism since gastrin, under some circumstances, can be a potent calcitonin secretagogue. It has also been suggested that production occurs by bone invaded by tumor of a substance to inhibit its resorption by stimulating calcitonin secretion. According to this theory, bone cells are able to produce a substance which can protect them from increased resorption [175]. The theory is attractive in that it suggests the autoregulation of bone turnover and metabolism. It is supported by the observation that increased bone resorption is a common factor among a variety of conditions associated with elevated plasma calcitonin. These include, in addition to malignancy with bony metastases, pregnancy, renal osteodystrophy, pancreatitis, and primary hyperparathyroidism. However, the concept of autoregulation of bone mediated by calcitonin secretion awaits experimental evaluation.

Artifactual Elevations in Plasma Calcitonin

While it is established that hypercalcitoninemia occurs in a variety of cancers, the immunologic measurement may reflect a nonspecific effect of the malignant state [156, 160, 162]. Abnormal immunoglobulins or other proteins, especially in myelomata and lymphomata, may interfere with the radioimmunoassay. Similarly, enzymes may nonspecifically affect either the calcitonin, the tracer, or the antibody used in the assay [156, 180, 492].

Various studies have supported each of these hypotheses regarding the pathogenesis of the hypercalcitoninemia of malignancy. These studies have demonstrated that (a) nonthyroidal malignancies produce calcitonin, (b) that tumor-associated hypercalcemia increases plasma calcitonin, (c) that tumor-associated factors can act as calcitonin secretagogues, (d) that calcitonin associated with malignancy exists in 'abnormal' forms, and (e) that calcitonin assays can be influenced by nonspecific factors in malignant disease [57, 135, 169–171, 173, 180, 255, 439, 458, 484].

Other Hypercalcitoninemic States

Hypercalcemia and Primary Hyperparathyroidism

Calcium challenge is a well documented stimulus for calcitonin secretion [175]. Increased calcitonin secretion has also been associated with chronic hypercalcemia and hypercalciuria [277, 321, 402]. This is presumed to be a homeostatic response by the C cells to defend against the challenge to calcium homeostasis. This abnormality may require provocative testing for its demonstration [402]. The relative calcitonin deficiency of females becomes manifest in primary hyperparathyroidism and may explain some aspects of the greater severity of bone disease in women with the disorder [195, 401, 402].

Hypocalcemia

The effect of calcium challenge on calcitonin secretion has been studied in hypocalcemic patients as well [165]. Calcium infusion (and to a lesser degree, pentagastrin) in patients with hypocalcemia of several etiologies results in abnormal increase of plasma calcitonin, presumably due to release of hypocalcemia-induced increased stores of the hormone [165].

Renal Disease

While there are increases in immunoassayable calcitonin with both acute and chronic renal failure [25, 31, 131, 252, 272, 322, 402, 440, 465], there is considerable disagreement regarding the significance of the increases. It has been reported that when patients with renal disease are dialyzed or receive calcium infusion, their already elevated calcitonin increases further [171, 252, 321, 400, 402, 440]. Other studies indicate no increase in these patients upon dialysis or calcium infusion [272, 466]. It is still not clear whether the elevation represents an abnormality of hormone secretion, metabolism, or both. However, since the kidney is active in the metabolism of the hormone, it is likely that the abnormality is metabolic, at least in part [23, 213]. This is supported by the fact that the metabolic clearance rate of calcitonin in renal failure is decreased [22]. Moreover, resolution of the renal failure is associated with a return to normal calcitonin levels [171, 252, 322]. Since the secretion and/or metabolism of calcitonin is abnormal in renal disease and since renal osteodystrophy is characterized by, among other features, increased bone resorption, then calcitonin, which acts to inhibit bone resorption, may be implicated in the pathogenesis of uremic osteodystrophy [131, 252]. In support of this view is the observation that lesser bone involvement in renal disease is associated with the greatest increases in plasma calcitonin which may be compensatory to the increased bone resorption [291].

Both hormone secretion and hormone metabolism may be involved in the mechanism of hypercalcitoninemia seen in renal failure. Since calcium levels are typically low in these patients, there is no obvious reason for increased secretion of calcitonin. However, serum gastrin is increased in renal failure and pentagastrin is a potent calcitonin secretagogue [309]. Additionally, other chronic hypergastrinemic states, such as pernicious anemia, are reported to have hypercalcitoninemia. However, in renal disease [253], there is no correlation between levels of calcitonin and plasma gastrin.

It has been noted that the calcitonin found in the plasma of patients with renal disease has a relatively small proportion of calcitonin monomer, unlike other hypercalcitoninemic states (such as MTC) in which the monomeric form of calcitonin seems to be dominant [321]. The dominant form of calcitonin in patients with renal disease is of a much higher molecular weight and unknown biological significance. There even appear to be differences between the species of calcitonin in the blood of patients with acute renal failure versus dialysis patients [321]. In the patient undergoing dialysis, monomeric and other small species of calcitonin may be lost through the dialysis, thus decreasing plasma-calcitonin concentration. Conversely, the calcium load of the dialysis may represent a challenge sufficient to stimulate all calcitonin species, with the larger less likely to be removed via dialysis.

It is possible that the abnormal immunochemical forms of calcitonin found in renal disease are less biologically active. In rats, activity as measured by radioimmunoassay is reduced and calcitonin response to calcium challenge is blunted when the animal tested has acute renal failure [322]. It may be that established renal failure is characterized by low levels of biologically active calcitonin with high levels of inactive hormone, while early renal failure might be characterized by calcitonin deficiency alone.

Bone Disease

Calcitonin secretion is abnormal in patients with pycnodysostosis, and may be altered in other hyperostotic states as well [41]. Reduced calcitonin reserve in females may account for the greater severity of osteitis fibrosa cystica in women with primary hyperparathyroidism [246, 401, 402]. If osteoporosis is due to elevated bone resorption, then calcitonin deficiency may play a role in this disease as well [49, 78, 104, 133, 366, 498]. This view is supported by studies in which calcitonin has been of therapeutic benefit [49, 78, 104, 133, 366]. Unfortunately, calcitonin is not consistently effective in cases of osteoporosis, leaving the issue still unresolved [430].

Pancreatitis

The pathogenesis of the hypocalcemia of pancreatitis has been attributed to increased calcitonin secretion, although not all studies have demonstrated calcitonin abnormalities in association with pancreatitis [103, 435].

In general, this is a controversial issue [35, 37, 103, 134, 317, 383, 398, 435, 527]. Studies in which elevated levels of calcitonin have been reported [103, 435] support the hypothesis that pancreatitis is characterized by the release of glucagon and that glucagon stimulates calcitonin secretion [35, 37, 317, 383, 398]. Alternatively, the pancreas or adjacent gastrointestinal cells may be a source of calcitonin-like activity [485, 497]. It is also possible that plasma artifacts may be exaggerated in pancreatitis, distorting radioimmunoassay findings [175, 492].

Production of Other Bioactive Substances by MTC

ACTH Production and Cushing's Syndrome

History

Since *Brown's* [92] description in 1928 of a case of diabetes and hirsutism in a patient with oat-cell carcinoma of the lung, a wide variety of tumors has been associated with ectopic ACTH production and Cushing's syndrome. In 1959, *Dyson* [202] added MTC to this syndrome by describing a patient with Cushing's syndrome and elevated urinary 17-ketosteroids who had a thyroid tumor which he classified as undifferentiated but which could retrospectively be diagnosed as MTC [541]. In 1967, *Goldberg and McNeil* [223] described a similar patient and demonstrated by bioassay that the thyroid tumor did, indeed, contain abnormal concentrations of ACTH. As in *Dyson's* [202] case, the tumor was correctly diagnosed as MTC, retrospectively. In 1968, *Donahower* et al. [197] described 2 patients with MTC and associated Cushing's syndrome; elevated ACTH was measured by bioassay in both tumor and plasma. Their review of the literature added additional cases to the syndrome [16, 260]. The association between MTC with ectopic ACTH production was emphasized by *Williams* et al. [541] in 1968 in a report of 2 new cases and a review of 9 additional cases in the literature. In the case reported by *Melvin* et al. [349], elevated ACTH and calcitonin in tumor and plasma were demonstrated by radioimmunoassay. *Croughs* et al. [147] demonstrated the presence of ACTH using immunohistological methods in the MTC of their patient with Cushing's syndrome. This provided evidence that the same cells were responsible for ACTH and calcitonin production. In other studies, however, the cells and, more importantly, the secretory granules containing the ACTH activity were different from those containing calcitonin activity [98, 285, 290]. It should be kept in mind that Cushing's syndrome can also occur with thyroid cancer which is not MTC [201, 260, 282, 299].

Pathogenesis

These studies have clearly established an association between MTC and Cushing's syndrome. It is estimated that 2–4% of the patients with MTC have Cushing's syndrome [541]. The most common pathogenesis of the Cushing's syndrome in the patient with MTC is the production of ACTH by the tumor. This view is additionally supported by the amelioration of the biochemical and clinical features of Cushing's when the thyroid tumor is removed [349]. However, alternative explanations are possible for some cases of Cushing's syndrome associated with MTC. In 1 patient reported by *Steiner* et al. [488], the findings were most consistent with Cushing's disease; the excessive level of ACTH was produced by a pituitary tumor. Another explanation to be considered is the ectopic production of ACTH or cortisol by one of the other tumors associated with MTC. ACTH has been demonstrated in several pheochromocytomas [325, 347]. Cushing's syndrome has been described in several patients with pheochromocytomas which were reported as nonfamilial. In some of these patients, adrenal hyperplasia was discovered at surgery. The Cushingoid features regressed following removal of the pheochromocytoma [347, 436]. Cushing's syndrome has also been reported in association with parathyroid disease and MTC [422, 436]. In 1 patient, resection of three hyperplastic parathyroid glands led to remission of the features of Cushing's syndrome [436]. The production by pheochromocytomas of cortisol has been demonstrated to occur in vitro [52], but this mechanism has not been implicated in the pathogenesis of Cushing's syndrome [52, 196].

There are also alternative explanations for the functional relationship between the adrenal cortex and MTC. The work of *Wurtman* [551] suggests that in some patients with excessive adrenal cortisol production, the high levels of cortisol production by the adrenal cortex may induce certain enzymes in the adrenal medulla. This process could lead to adrenal medullary hyperplasia, the predecessor of MTC in some patients. This must be regarded as interesting speculation which currently has no demonstrated relevance for the hypercortisolism seen in MTC. However, these observations have clinical value. ACTH administration in patients with pheochromocytoma may precipitate an adrenergic crisis [373]. This may be due to increased catecholamine secretion induced by the rapid increase in the intra-adrenal cortisol concentration and the subsequent induction of adrenal medullary enzymes which regulate catecholamine metabolism. The validity of this explanation notwithstanding, great caution should be exercised when considering the diagnostic use of ACTH in patients with a potential pheochromocytoma.

Clinical Features

Tumors which ectopically produce ACTH are usually rapidly progressive and highly lethal. The patient does not usually survive long enough to develop the classical physical findings of Cushing's disease, such as moon faces, hirsutism, and striae. The clinical course is often dominated by the tumor itself and by the biochemical abnormalities of hypercortisolism. In fact, the presence of ectopic ACTH production is often suggested by the biochemical abnormality, hypokalemic alkalosis, rather than by physical findings [326].

Although some MTCs also have a rapidly progressive clinical course, many of the tumors have more indolent behavior. Because of this, the patient with ectopic ACTH production by MTC may develop the same characteristics of Cushing's syndrome along with the biochemical abnormalities. In a review of this problem, *Rosenberg* et al. [443] evaluated the clinical and biochemical features of 10 patients from the literature and 1 of their own patients with ectopic ACTH production by an MTC. In 6 cases, the tumor was sufficiently indolent so that the patients developed the typical manifestations of Cushing's syndrome. In the other 4 cases, the tumor had a more rapid course and the clinical features of Cushing's syndrome were less prominent.

Histaminase

Elevated levels of histaminase, an enzyme which catalyzes the deamination of histamine and several other polyamines, are commonly found in tumor and serum of patients with MTC [19, 52–54, 56]. Histaminase is increased in the primary as well as metastatic tumor. *Baylin* [59] and *Baylin* et al. [54] found abnormal histaminase levels in the serum of 50% of patients with MTC. 70% of the elevations were found in patients with metastatic disease and the highest elevations were in the same group. Aminoguanidine, a histaminase inhibitor, has been administered to MTC patients with elevated histaminase levels. This has resulted in a decrease in serum histaminase but no apparent effect on the secretion of calcitonin by the tumor or its progression [52, 352]. High levels of histaminase were specific for MTC, and not found in other tissues, including associated pheochromocytomas, or in a variety of other tumors [52–54, 56]. This finding is in contrast to prostaglandin levels, which are also increased in MTC as well as in other cancers.

In addition to the fundamental importance of this observation, the elevated histaminase levels may have some clinical utility. Histaminase is generally not as specific as calcitonin in identifying patients with MTC. Its concentration does not correlate as well with surgical removal of the tumor [56]. However, in some patients histaminase measurements may be more specific than calcitonin measurements in identifying those with residual tumor after treatment [56]. In one group of 11 patients with MTC, 7 patients had increased serum histaminase and all patients had increased serum calcitonin preoperatively [54, 56]. In 5 of the 7 patients with increased histaminase, the enzyme level remained elevated postoperatively, despite the return to normal of basal and stimulated calcitonin levels. 2 of these 5 patients eventually demonstrated residual tumor with abnormal calcitonin values. Despite these findings, histaminase is not as reliable an index of MTC as calcitonin. However, histaminase measurements may help to distinguish between CCH and MTC, since elevated levels are measured only in the latter [55].

The elevated histaminase levels may play some role in the abnormal intradermal histamine test observed in patients with MTC in which injection of histamine produces a wheal but no flare [50, 210, 229, 303] by inactivating the histamine. Alternative explanations must be considered for this phenomenon since it can occur in some patients with MTC who have normal histamine levels [92, 210]. It can also be present in patients with pheochromocytomas not apparently associated with MTC and it can return to normal after tumor removal. Because of these observations, and since an abnormal histamine test can also be seen in certain neurological disorders, it has been postulated that the neural abnormalities seen in patients with multiple endocrine neoplasia (MEN) type IIb may contribute to their abnormal histamine flare [92, 303].

Prostaglandins

In 1968, *Williams* et al. [540] demonstrated elevated prostaglandin E1 and F2 levels in the tumor of 4 of 7 patients with MTC and in the blood of 2 patients; in these 2, the concentrations of prostaglandins in the blood draining from the tumor were higher than in peripheral blood. *Williams* et al. [540] suggested that the excess prostaglandins may be responsible for the commonly occurring diarrhea seen in patients with MTC. This original observation has been confirmed and now seems well established [68, 254,

274]. In addition, prostaglandins of the ABE series have also been implicated in the diarrhea of MTC [6, 350].

There is additional evidence to support the view that prostaglandin production by MTC contributes to the diarrhea commonly seen in patients with this tumor. Diarrhea seems to be a more prominent symptom in patients with extensive MTC and it seems to become less severe after surgical removal of the tumor. Prostaglandins can stimulate intestinal smooth muscle and their administration, either orally or parenterally, can produce diarrhea [149, 264, 344, 370]. Prostaglandin inhibitors, aspirin and indomethacin, seem to decrease the diarrhea in some patients with MTC [44]. Prostaglandins can be stimulated following alcohol ingestion, thus providing three mechanisms for alcohol-associated flushing in MTC-calcitonin, alcohol, and prostaglandins [132].

The relationship between diarrhea in MTC and excessive prostaglandin production by the tumor is not always straightforward. Patients with diarrhea and MTC have been reported to demonstrate normal prostaglandin levels [68, 254, 356]. In some patients with elevated prostaglandin levels, the highest levels are not always associated with diarrhea [541], and prostaglandin inhibitors are not always effective in treatment [274, 356]. Abnormal concentrations of prostaglandins are associated with a wide variety of tumors not associated with diarrhea [458]. Furthermore, prostaglandins are rapidly cleared from blood and it is not certain that the amount of prostaglandins produced by MTC is sufficient to produce diarrhea. Finally, other humoral and anatomical factors contribute to diarrhea in MTC. Therefore, prostaglandins must be considered as one of several potential etiological agents in the diarrhea of MTC.

Dopa Decarboxylase and Biogenic Amines

Dopa decarboxylase is an enzyme important for the synthesis of catecholamines and is located in cells and organs, such as the adrenal, where such synthesis occurs. Normal C cells from a variety of mammalian species contain this enzyme [35, 410]. This finding further supports the interrelationship of these cells with other neural-crest cells. *Atkins* et al. [29] reported that dopa decarboxylase is also found in MTC and in concentrations comparable to those seen in pheochromocytomas. Catecholamines and other biogenic amines are also present in MTC [190, 517].

Serotonin and Related Factors

Serotonin production by MTC has been implicated in the pathogenesis of the diarrhea commonly seen in this disease [371, 535]. Evidence has also been provided for abnormalities of kallikrein and bradykinin [62, 540]. It is interesting to note that carcinoid tumors can also produce abnormal concentrations of calcitonin [174, 364].

Carcinoembryonic Antigen (CEA)

The association between CEA and MTC was reported in 1976 by *Ishikawa and Hamada* [275]. Since then, several laboratories have demonstrated the presence by immunochemical means of CEA in the tumor and blood of patients with MTC [100, 102, 203, 530]. *DeLellis* et al. [194] furthermore demonstrated the presence of CEA in neoplastic C cells and its absence in hyperplastic C cells. In most studies, there is no consistent relationship between the blood concentrations of CEA and calcitonin. However, the presence of an elevated CEA is generally associated with progression or recurrence of disease. Although CEA measurements may thus provide some useful information about the management of the patient with MTC, this determination suffers from lack of specificity for MTC and is, by itself, not as valuable as calcitonin measurement.

Nerve-Growth Factor

Williams [542] speculated that MTC may produce a factor which stimulates the growth of nervous tissue. This view is supported by the occurrence of proliferation of neural tissue in MEN IIb (III) and by the more tenuous association with MTC of several central-nervous-system tumors [302]. Several studies have provided direct evidence to support this view by demonstrating the presence of nerve-growth factor in MTC [71, 146].

Other Peptides

Several other pituitary-hypothalamic peptides have been identified in MTC. These are somatostatin [183, 442, 494, 552], beta-endorphin [264], substance P [481], neurotensin [553], and beta-MSH [2]. Somatostatin is also present in normal C cells and can inhibit the secretion of calcitonin [545]. Abnormal concentrations of vasoactive intestinal peptide have been observed in patients with MMN [509]. Abnormal concentrations of calcitonin have been observed in vipomas [423]. *Birgenhager* et al. [73] also provided indirect evidence for production by MTC of releasing factors for ACTH and prolactin. These observations may establish a relationship for the neural-crest origin of the tumors of MEN II.

Other Substances

Several features of MEN II are likely to have a hormonal basis although no hormonal cause has yet been established. These include gynecomastia, erythremia and polycythemia vera, a desmoplastic reaction, and diabetes mellitus [254, 302, 333, 384, 517, 535].

Clinical Consequences of Hypercalcitoninemia

Biological Effects of Calcitonin

Bone, kidney, and the gastrointestinal tract are the three main target organs of calcitonin. As with other peptide hormones, calcitonin seems to mediate its effects via adenylate cyclase [107, 247, 343].

Bone

The major biological effect of calcitonin is to decrease bone resorption by inhibiting the activity of osteoclasts and perhaps the osteocytes [262]. This inhibition is reflected by the decrease in bone gla protein (BGP) and in alkaline phosphatase produced by calcitonin [184]. Calcitonin may also promote bone formation, but this has not been firmly established [526]. Calcitonin acts on both the mineral and organic phases of bone. Calcitonin prevents the translocation of calcium and phosphorus from bone by inhibiting its resorption. In a situation in which bone turnover is sufficiently high, calcitonin will produce hypocalcemia and hypophosphatemia. Calcitonin decreases the excretion of urinary hydroxyproline by inhibiting the resorption of the organic phase of bone [259].

Kidney

Urinary excretion of calcium and phosphorus is increased in response to calcitonin [20, 21, 24]. Renal excretion of sodium and chloride, and to a lesser extent potassium, is also promoted by calcitonin [72]. Since studies of these effects utilize pharmacological doses of calcitonin, the physiological significance of these renal effects of the hormone are not fully understood [25, 32].

Gastrointestinal Absorption of Calcium

The effect of calcitonin on the gastrointestinal tract is not firmly established [235]. In vivo rat studies by *Milhaud and Moukhtar* [361] demonstrated that the administration of calcitonin produced an increase in intestinal calcium absorption and a decrease in endogenous fecal calcium excretion. However, the study did not exclude parathyroid hormone as the indirect mediator of this effect. *Olson* et al. [392] studied the effect of calcitonin on isolated rat intestinal segments perfused with an artificial blood supply. They found that small doses of calcitonin produced a marked and rapid increase in calcium absorption. This increase occurred more rapidly than the rise observed after administration of parathyroid hormone. This study suggested that parathyroid hormone did not mediate the calcitonin effect. Calcium absorption was not effected on isolated gut loops in dogs by hog or chicken calcitonin [144]. *Cramer* [145] found no effects of salmon calcitonin on calcium absorption in rats that were repeatedly injected with the hormone over several hours.

Other Effects

The effects of calcitonin have been observed to be widespread. It has been reported to act as an anti-inflammatory agent [3, 514], to promote fracture healing [4] and wound healing [336], and to be uricosuric [75]. It has been observed to act as an antihypertensive, an analgesic, and a lipolysis inhibitor [9, 86, 531]. It has also been reported to impair glucose tolerance [76] and to exert effects on the pituitary [323] and the central nervous system [413]. The importance of these effects is yet to be determined.

Role of Calcitonin in Mineral Metabolism

The physiological role of calcitonin with respect to calcium homeostasis or skeletal metabolism has not been established in humans. Athyreotic adults do not seem to have a clearly defined abnormality in skeletal metabolism as assessed by short-term studies. However, these studies should be interpreted with reservation. For example, it is difficult to determine if adults who have been thyroidectomized are truly calcitonin-deficient [74]. Thyroidectomy or ablation with radioiodine may leave remnants of thyroid tissue. There may also be extrathyroidal sources of calcitonin [179]. Both plasma and urinary calcitonin have been described in presumably athyreotic adults [471]. In addition, the role of a potentially damaged parathyroid has not been carefully considered in such studies, nor have there been long-term studies conducted to establish the presence or absence of mineral abnormalities in athyreotic adults. Bone disease is common among athyreotic cretins, but it is generally ascribed to deficiencies of T3 and T4. However, a role for calcitonin deficiency in the pathogenesis of thyroid-related bone disease has not been ruled out [111].

Calcitonin may play a role in blood-calcium homeostasis. Some thyroidectomized patients demonstrate a decreased tolerance to oral and parenteral calcium challenge, and patients with MTC and high levels of calcitonin demonstrate an increased calcium tolerance [349]. In animals, feeding increases, and fasting decreases plasma calcitonin [441]. These and other observations in animals notwithstanding, human studies have failed to demonstrate a postprandial increase in plasma calcitonin which cannot be accounted for by an increase in blood calcium [34].

Abnormalities in calcium or bone have not been consistently found in patients with MTC characterized by increased plasma calcitonin. However, it is possible that the escape of bone from the biological effect of calcitonin has minimized such changes, perhaps in a homeostatically appropriate manner [421]. It remains possible that an increase in parathyroid hormone secretion in such patients, either in a compensatory manner or due to genetic factors, obviates the effect of calcitonin. There is even evidence in very careful studies of a presumed calcitonin effect on bone in MTC patients [353]. The hypocalcemic effect of calcitonin appears to decrease with age, according to animal studies. Thus, it has been suggested that calcitonin plays a role in the growth and development of bone [175]. Elevated levels of plasma calcitonin in neonates tend to support this view [448].

The regulatory role of calcitonin in calcium homeostasis and metabolism is not entirely clear. In normal animals, the elimination of calcitonin does not lead to a rise in steady-state blood-calcium levels, nor does it affect the steady-state parameters of calcium metabolism. Studies utilizing an intraperitoneally injected calcium load revealed that in calcitonin-deficient (thyroidectomized with functioning parathyroid autografts) animals, it took longer than in controls for the plasma-calcium level to return to the preinjection level. Thus, the ability to correct an acute positive error in plasma-calcium regulation is diminished in animals without circulating calcitonin. Since it is difficult to induce acute hypercalcemia by nutritional means, even in calcitonin-deficient animals, the physiological significance of this is not known. An increase in plasma calcium is seen in thyroidectomized older rats, who have relatively high levels of plasma calcitonin. The observed elevation is, however, slight and transient [289].

Pregnancy, suckling, and lactation result in an increase in plasma calcitonin in rats [138, 503, 508]. This may protect the fetus and neonate against postprandial hypercalcemia and promote the assimilation of calcium into the developing skeleton. The elevated calcitonin may also attenuate any increased bone resorption that occurs to provide calcium in milk (lactation), thus benefiting the mother as well. In general, in humans as well as bovines, basal plasma calcitonin is lower in females than in males, and females have decreased calcitonin reserve during provocative testing [179, 478]. In bovines, a high calcium diet can bring female plasma-calcitonin levels up to male levels [164]. This disparity associated with gender may play some role in the pathogenesis of bone diseases in females. Females with primary hyperparathyroidism are unable to increase their plasma calcitonin in the face of this hypercalcemic (and bone hyperresorptive) challenge while hyperparathyroid males can do so [402]. This may explain the higher incidence of bone disease found in hyperparathyroid females [397]. Calcitonin deficiency may also be involved in the pathogenesis of postmenopausal osteoporosis since calcitonin secretion in females decreases with increasing age [185].

It has been proposed that immunoreactive calcitonin, or a calcitonin-like substance identified in the pituitary gland of several mammalian species [177, 178], may perform a paracrine as well as an endocrine function. This is particularly interesting taken in the context of menstruation-related changes in plasma calcitonin in normal females [416], and the variety of local, often inhibitory, effects with which calcitonin is often associated.

While a number of hypotheses are available regarding the physiological significance of calcitonin, and a variety of techniques are now available to monitor the hormone, a precise role has yet to be defined. New functions continue to emerge, and the understanding of the spectrum of actions must await further explorations [33, 176].

Newer Concepts – Calcitonin as a Neurotransmitter

Calcitonin may have multiple paracrine functions in addition to the endocrine activities described above. Calcitonin or a calcitonin-like peptide has been identified in a variety of normal and malignant cells, including the pituitary, pancreas, gastrointestinal tract, thymus, lungs, testes, adrenals, ovaries, and the parathyroids [65, 77, 123, 140, 178, 186, 191, 276, 523]. A variety of paracrine substances, including somatostatin, substance P, and beta-endorphin have similar distributions [12, 187, 188, 461, 553]. These other paracrine substances function as neurotransmitters. In addition, there are a variety of processes and sites upon which calcitonin has been demonstrated to exert an inhibitory action [15, 105, 125, 126, 155, 216, 266, 273, 304, 307, 393, 413, 415, 493, 531], including the lung, heart, gastrointestinal tract, gall bladder, salivary gland, pancreas, pituitary, and brain. While calcitonin may act as an endocrine hormone at some of these sites, it may be actually synthesized in many of these same tissues, as well. Thus, calcitonin may have a general paracrine function of inhibiting cell function. This may represent a manifestation of calcitonin's role in biological communication. In unicellular organisms, hormones act as intracellular regulators; in oligocellular organisms, hormones can act as a paracrine mediator of cell-to-cell communication; in multicellular organisms, hormones have an endocrine effect. Applying this developmental sequence to calcitonin, it may be that in neural tissue calcitonin is a neurotransmitter [217, 218], and in more complex organisms calcitonin becomes an endocrine hormone [174]. While this evolutionary sequence might explain the widespread distribution of calcitonin, its paracrine, neurocrine, and endocrine functions may still be mediated by the same mechanism – the translocation of calcium and phosphate, either intracellularly or intercellularly [85].

The Patient with MTC

The biological effects of the abnormal concentrations of calcitonin in MTC may not be dramatic. There are several possible explanations for this phenomenon. A most obvious possibility is that the calcitonin produced by MTC is not biologically active. This hypothesis is not tenable since the biological activity of calcitonin isolated from MTC has been conclusively demonstrated. Furthermore, structurally identical synthetic human calcitonin is biologically active. Another possibility is that the biological activity of calcitonin is counteracted by the presence of its antagonist, PTH. This explanation has some merit. A significant percentage of patients with MTC do have high levels of PTH which can block the action of calcitonin at several organ sites. However, even patients with MTC and normal levels of PTH do not seem to exhibit any effects that might be expected of calcitonin excess. A most attractive explanation for the apparent lack of calcitonin exposure in patients with MTC has been suggested by the work of *Raisz* et al. [421] and of *Tashjian* et al. [502]. They observed that in vitro preparations of bone cells which were continually exposed to calcitonin became unresponsive to the hormone. This was not entirely due to a progressive loss of biological activity of the incubating hormone. *Tashjian* et al. [502] extended these results by demonstrating a decrease in the number of receptors for calcitonin on the bone cells. These observations suggest that the lack of a prominent calcitonin response in patients with MTC may be the result of down modulation at the receptor level to maintain homeostasis. If such a receptor mechanism exists, it represents an additional homeostatic system for regulating hormone action.

The subtlety of biological effects of calcitonin notwithstanding, the patient with MTC can present to the physician in a variety of ways. Since the tumor is relatively uncommon, the physician must be aware of its protean manifestations in order to make the correct diagnosis in the patient and to additionally consider the diagnosis in the patient's relatives. It is important to keep in mind that the patient with MTC may have no signs or symptoms of the tumor. The patient may be asymptomatic but referred to the physician because a relative was found to have MTC. The following symptoms and signs should suggest the possibility of MTC. The most common sign is an enlargement in the thyroid gland and the most common symptom is diarrhea.

Bone Disease

Abnormalities of bone are seldom detected in patients with MTC by the usual clinical tools. Except for the presence of metastases, bone X rays are usually normal as are plasma alkaline phosphatase and urinary hydroxyproline, indices of bone formation and resorption, respectively. Balance and kinetic studies of bone metabolism are also normal. The lack of bone changes has been surprising when one considers the high levels in these patients of plasma calcitonin which is biologically active. Several explanations can be offered for the apparent lack of inhibition of bone resorption by the high levels of the hormone. In some patients, the potential effect of calcitonin may be countered by associated hyperparathyroidism and high levels of parathyroid hormone, the physiological antagonist to calcitonin on bone. Another possible explanation for the lack of a calcitonin effect on bone in patients with MTC may be due to a decrease in the number of receptors for the hormone [421, 502].

There are, however, bony changes in some patients which may be due to the abnormal concentration of plasma calcitonin. Despite the lack of X-ray changes, bone biopsy in affected patients does reveal a decrease in the number of osteoclasts and a decrease in osteocytic osteolysis [353]. More dramatic than these subtle histological changes were the observations made by *Verdy* et al. [515]. They studied a patient with MTC over the course of a 5-year period. During that time, the patient's skeletal survey changed from normal to a picture of abnormally dense bones. However, calcitonin excess was not the only explanation for these X-ray changes. The patient also gave birth to 4 children who also had dense bones. 2 of the children had normal plasma-calcitonin concentrations. These observations suggest that the abnormally dense bones could have been a manifestation of a genetic bone disease rather than calcitonin excess. These dramatic findings notwithstanding, most patients with MTC do not have bone abnormalities unless caused by tumor metastases.

Blood and Urinary Minerals

The concentration of calcium and phosphate in peripheral blood is normal in most patients with MTC. This finding may be surprising since the administration of calcitonin can produce hypocalcemia and hypophosphatemia. Decreased calcitonin receptors and/or a compensating increase in PTH may explain these blood chemistries which usually occur in MTC. There are, however, some studies which do suggest altered mineral metabolism in MTC. *Miravet* et al. [369] observed a resistance to the effects of parathyroid extract and 25(OH)vitamin D in MTC. *Paterson* [403] similarly reported resistance to the effect of vitamin D in a patient with hypoparathyroidism and MTC. There is experimental evidence that calcitonin inhibits the formation of 1,25(OH)2D [426] and promotes the formation of 24,25(OH)2D [37]. 25(OH)D3 has been reported to interfere with the phosphaturic effect of calcitonin [418], and MTC can be characterized by a slight hyperphosphatemia [425]. Magnesium is normal [425]. Some patients with MTC may be slightly hypocalcemic and demonstrate greater tolerance to calcium challenge [352]. However, the hypocalcemia may be due to a commonly accompanying diarrhea [1].

Since calcitonin is a potent natriuretic and calciuretic agent, it might be expected to produce this effect in patients with MTC, but the renal excretion of electrolytes is normal in patients with this tumor. However, *Krane* et al. [311] have demonstrated that the kidneys in patients with MTC are responsive to further doses of calcitonin. They observed that the administration of salmon calcitonin to such patients produced an increase in the urinary excretion of calcium. This effect was contrasted with the absence of a decrease in urinary hydroxyproline which signified the lack of a bone effect of the hormone. Thus, in patients with MTC, the kidney maintains its responsiveness to calcitonin whereas bone may not.

Kidney Stones

Kidney stones seem to occur with an increased frequency in patients with MTC. This is usually explained by the common association with MTC of hyperparathyroidism and the consequent hypercalcemia and hypercalciuria [127]. However, there is some clinical evidence to suggest that the increased calcitonin may result in the hyperabsorption of calcium from the gastrointestinal tract and that this abnormality is responsible for the nephrolithiasis [84]. Experimental studies of the effect of calcitonin on the absorption of calcium have produced inconsistent and inconclusive results [144, 145, 367, 392, 393]. However, clinical studies in patients with MTC demonstrate increased calcium absorption. When the tumor is removed, calcium absorption returns toward normal. These events occur independently of any changes in parathyroid hormone or blood calcium. Although additional studies are necessary to confirm this pathophysiological consequence of hypercalcitoninemia, the current data suggest that kidney stones may be another feature of MTC due to the abnormal concentrations of calcitonin [127].

Diarrhea

Diarrhea is one of the most common symptoms in patients with MTC. It occurs in approximately one-third of patients with the tumor, and it can even precede the diagnosis [68, 254]. The diarrhea is characterized by a rapid transit time of both the large and small intestine [68, 293]. This may produce a diabetic glucose tolerance test [68]. There is excessive loss of fluids and electrolytes due to their poor absorption [274]. These abnormalities occur primarily in the ileum, and jejunal function seems to be normal [274]. Malabsorption, as evaluated by B_{12} and xylose absorption, does not occur, and steatorrhea is mild or absent; however, cellulose and starch are apparent in the stool [68]. Intestinal biopsy is normal or mildly abnormal; in the latter circumstance there is mild inflammation and some villous atrophy, but the epithelium is usually normal. However, the radiological appearance of the gastrointestinal tract can sometimes be confused with ulcerative or granulomatous colitis [412, 521].

There are 2 general groups of factors which can contribute to the diarrhea commonly seen in patients with MTC – humoral factors and anatomical factors. Many of the various bioactive substances produced by MTC have been implicated in the pathogenesis of the diarrhea seen with this tumor. Such peptide hormones are: (1) calcitonin; (2) ACTH and MSH; (3) neurotensin; (4) somatostatin; (5) beta-endorphin; (6) nerve-growth factor. The reasons these substances have been implicated in the etiology of diarrhea can be summarized as follows: (a) the substances can produce diarrhea when administered at the high levels found in patients with MTC; (b) diarrhea can be ameliorated by the administration of specific inhibitors of the substances; (c) removal of the tumor and the concomitant decrease in blood concentrations of the substances can be accompanied by a decrease in the diarrhea; (d) the extent of diarrhea can be roughly proportional to tumor size and can parallel tumor recurrence; and (e) diarrhea can occur in the absence of any demonstrable (usually by X ray) organic lesion of the intestine. However, the relationship between diarrhea and these various agents is not always direct and in many patients other causes must be sought for the diarrhea.

The various anatomical abnormalities discussed subsequently which can be found in the gastrointestinal tract of patients with MTC may also account for the diarrhea. These anatomical lesions can reflect and perhaps even produce fundamental abnormalities in gastrointestinal innervation that can produce abnormal motility.

The diarrhea of MTC may respond, at least partially, to standard antidiarrheal regimens of atropine-like agents [68, 274, 293]. Removal of the tumor may decrease the diarrhea; the presumed mechanism is a decrease in the concentration of a humor responsible for the diarrhea. Specific antihumoral substances, such as aspirin and indomethacin for prostaglandins may be beneficial in some patients. If alcohol ingestion accentuates the diarrhea, it should be eliminated. If an anatomical lesion, such as ganglioneuromatosis, is thought to contribute to the diarrhea, surgery is not likely to be effective since the lesion is likely to be diffusely distributed.

There are some anecdotal reports of the value of nutmeg *(Myristica fragrans)* in the treatment of the diarrhea of MTC [44, 208]. Nutmeg is a mixture of volatile oils, fats, myristin, elemecin, and safrale, and may act in several ways. It may be antidiarrheal because of the atropine-like properties of myristin [44], or the sympathomimetic effects of elemecin [456]. Although its use should be considered in the treatment of diarrhea associated with MTC, nutmeg seems to offer no advantage over more standardized antidiarrheal agents. Interestingly, nutmeg has also been reported to decrease hypercalcemia [274].

Peptic Ulcer Disease

There is some evidence to suggest that there may be an increased incidence of peptic ulcer disease in MTC. *Hill* et al. [303] observed that 5 of 44 patients with MTC had peptic ulcer disease and 2 of them had multiple ulcers. They reiterated the relationship proposed by *Pearse* [405, 408], *Pearse* et al. [410], and *Weichert* [525] between ulcerogenic tumors [204] and MTC and its associated tumors. It was suggested that MTC may produce an ulcerogenic factor. However, only 1 patient with MTC and increased plasma gastrin has been reported [274]. The majority of patients showed decreased plasma gastrin due to feedback inhibition caused by calcitonin [477]. *Ljungberg* [333] and *Walker* [520] also reported the occurrence of ulcers in patients with MTC [520]. The increased incidence of ulcer disease may also be due to hyperparathyroidism associated with MTC. It is not possible to determine from existing data if ulcer disease is directly associated with MTC [251].

Carcinoid Syndrome

In a review of the wide variety of tumors associated with the carcinoid syndrome – flushing, abdominal pain, and diarrhea – *Moertel* et al. [371] reported a case of MTC with increased serotonin in blood and tumor and increased 5-HIAA in blood. In 1966, *Williams* [535] confirmed this finding and suggested that serotonin excess might be a mechanism for diarrhea in MTC. Subsequent case reports have further established the association between MTC and the carcinoid syndrome. In some patients the carcinoid syndrome can be precipitated and even uncovered by the ingestion of alcohol [132] or the infusion of calcium, perhaps by stimulating plasma calcitonin [295]. Production by the tumor of substances integral to the carcinoid syndrome, such as bradykinin [62] and kallikrein [540], may also play a role in the carcinoid symptomatology.

The carcinoid syndrome can also occur with other tumors of neural-crest origin. These tumors are, therefore, embryologically related to MTC. Tumors of neural-crest origin have been reported to produce calcitonin. Further links between MTC and the carcinoid tumor have been reported: both MTC and carcinoid tumor have been found in the same patients, there is a histological similarity between the two tumors, and carcinoid tumors produce calcitonin [173, 191, 294, 352, 363].

Although it is well established that the classical biochemical and clinical features of the carcinoid syndrome can be produced by MTC, the symptoms of some patients with MTC may not be ascribed to increased serotonin. In a series of MTC [254], 1 patient had carcinoid symptoms with normal plasma serotonin, 11 patients had facial flushing only, and urinary 5-HIAA levels were normal in the 5 patients measured. These findings suggest that, as with diarrhea, some of the symptoms of the carcinoid syndrome may actually be due to other substances produced by MTC. Carcinoid-like symptoms can be produced by the administration of prostaglandins [264], pentagastrin [249, 274], and even calcitonin [234, 548, 549], each of which can be increased in patients with MTC.

There is a significant incidence of carcinoid tumors in patients with MTC [200, 363]. The occurrence of carcinoid tumors should be noted for the following reasons: (a) the cells of carcinoid tumors and MTC are of neural-crest origin; (b) the histological appearance of MTC can resemble carcinoid tumors; (c) carcinoid tumors can produce calcitonin, and (d) MTC can produce serotonin. These observations provide additional support for the embryological, histological, and functional relationship of the tumors which occur in MEN.

Hypertension

The presence of pheochromocytomas in MTC can account for hypertension in these patients. However, the clinical and biochemical features of the hypertension may differ from those observed in patients with nonfamilial pheochromocytomas. The hypertension may not be sustained and the routine tests may be nondiagnostic.

Pigmentation

Abnormal pigmentation has been described in several patients with MTC [45, 151, 283, 302, 333, 396, 445]. In one study [151], a patient presented with melanin pigmentation around the mouth, joints, and on the dorsal aspects of the toes, hands, and fingers; all completely disappeared following removal of the tumor. It is possible that a pigmentary substance, perhaps related to ACTH, can be added to the list of bioactive substances produced by MTC.

Gynecomastia

Gynecomastia has been reported in 2 patients with MTC [1, 352]. Although no humoral agent has been implicated in the gynecomastia, such a finding might be anticipated.

Marfanoid Habitus

The presence of a Marfanoid habitus along with MMN is essentially diagnostic of familial MTC. The distinction from Marfan's disease can usually be made on clinical grounds.

Mucosal Neuromas

Mucosal neuromas are found in only a minority of patients with MTC. Their importance as clinical manifestations of the presence of MTC derives from the fact that they are often early manifestations of the MMN syndrome. They usually occur within the first decade and may even be present at birth. They can represent early warning signals for the potential presence of MTC and pheochromocytoma. If the mucosal neuromas alert the physician to the early diagnosis of these two tumors, early and potentially curative therapy can be instituted. Equally important, the patient's family can also be screened for the other components of MEN. No carotid-body tumors or paragangliomas have yet been reported in association with MTC.

Cushing's Syndrome

A small percentage of patients with MTC exhibit the ectopic production of ACTH by the tumor. This can produce the rapidly progressive Cushing's syndrome associated with ectopic ACTH production. Hypokalemic alkalosis is the dominant feature of this type of Cushing's syndrome since the clinical course is not long enough for the more classical somatic features of Cushing's syndrome to develop. However, some MTCs which ectopically secrete ACTH run a more indolent course. In such patients, they may develop the more classical picture of Cushing's syndrome.

Cervical Enlargement

The discovery by the patient of a mass in the anterior neck is the most common presentation of the patient with MTC [254]. The mass usually represents a single thyroid nodule which is cold on scan. However, patients with MTC have a higher percentage of multiple nodules and diffuse enlargement of the thyroid than patients with other thyroid cancers. The size of the mass may vary considerably. The characteristics of the tumor on physical examination are not diagnostic but some features may be more suggestive of MTC. The tumor may be more commonly palpated in the middle and upper third of the lobe of the gland and it is frequently demarcated and mobile [220].

Palpable regional lymph-node metastases are commonly present in patients with MTC at the time of diagnosis [127]. The patient may notice the cervical enlargement caused by lymph-node metastases before he notices the thyroid mass. Thyroid and lymphatic enlargement do not usually cause local symptoms but some patients may complain of a pressure sensation in the neck, dysphagia, hoarseness, and even vocal-cord paralysis [200, 254]. In one report the tumor caused fatal asphyxia [39].

Multiple Endocrine Neoplasia (MEN)

A most important clinical feature of MTC is its familial association with other endocrine tumors. MTC is part of an MEN syndrome which can be distinguished from MEN I. MTC occurs in association with pheochromocytomas, mucosal neural tumors (mucosal neuromas), and hyperparathyroidism. When it occurs with these other endocrine tumors, two distinct patterns can be defined. One of them, MEN II (or IIa) consists of MTC, pheochromocytomas, and hyperparathyroidism. The other, MEN IIb (or III), consists of MTC, pheochromocytomas, and neural tumors, with a much lower, if any, incidence of hyperparathyroidism; these patients can additionally have a Marfanoid habitus and the syndrome has been named MMN.

The association of MTC with these other tumors has an embryological as well as genetic basis. The cells of MTC, pheochromocytomas, and the neural tumors are all of neural-crest origin. Some authorities have even suggested a neural-crest origin for the parathyroid gland. An alternative explanation for the hyperparathyroidism is a functional relationship between it and MTC. According to this hypothesis, the abnormal concentrations of calcitonin produce hyperparathyroidism secondary to the hypocalcemic actions of the calcitonin. Although this type of functional relationship between the neoplasias may exist, the most convincing evidence supports a genetic relationship between MTC and hyperparathyroidism.

In addition to the fundamental importance of the multiple tumors associated with MTC, these associations have important clinical consequences. When they occur, the neural tumors may be the first manifestation of the MEN. Therefore, these lesions may provide an early warning to the presence of two potentially lethal tumors – MTC and pheochromocytoma. Additionally, the existence of pheochromocytoma or MTC should suggest the possibility of the coexistence of the other tumor. Hyperparathyroidism can be similarly regarded.

It is therefore important for the physician to recognize that the presence of MTC should stimulate a search for other tumors in this patient and a search for similar tumors in the patient's family. If diagnosed early, all of the serious features of MEN are treatable and even curable (table I).

Table I. Endocrine and nonendocrine neoplasms associated with MEN IIa and IIb

MEN IIa	MEN IIb
Medullary thyroid carcinoma	Medullary thyroid carcinoma
Pheochromocytoma	Pheochromocytoma
Hyperparathyroidism	Multiple mucosal neuromas
	Marfanoid habitus

Pheochromocytoma

History

In 1961, *Sipple* [476] presented evidence which supported the association between pheochromocytoma and thyroid tumor. He reported the case of a 33-year-old male with bilateral pheochromocytoma and a poorly differentiated and invasive thyroid tumor which was thought to be a follicular adenocarcinoma. This patient also had an enlarged parathyroid gland consistent with adenoma and a cerebral arteriovenous malformation. *Sipple* [476] reviewed the literature and presented 5 other cases with pheochromocytoma and thyroid tumor. The pheochromocytomas were bilateral in 4 of the cases and the thyroid tumors were variously described as follicular adenocarcinoma, papillary adenocarcinoma, adenocarcinoma and anaplastic carcinoma. *Sipple* [476] thus documented the previous speculation of *DeCourcy and DeCourcy* [158] and concluded that in patients with pheochromocytomas 'the incidence of carcinoma of the thyroid gland is increased far beyond expectation based on chance occurrence'. However, *Sipple* [476] did not appreciate the distinct features of MTC which had been described 2 years earlier by *Hazard* et al. [242]. In the early 1960s, there were additional case reports of the simultaneous occurrence of pheochromocytomas and thyroid cancer [209, 241, 340, 389, 395, 489]. Although this association between pheochromocytoma and MTC (and hyperparathyroidism) has come to be known as Sipple's syndrome or MEN II, it was *Williams* [534] who first appreciated this relationship. In a review of 15 cases along with the reporting of 2 of his own, he was able to establish that at least 11 of the total of 17 cases of thyroid tumor were actually MTC. In the same year, *Schimke and Hartmann* [453] also reviewed the previous reports of the simultaneous occurrence of pheochromocytoma and MTC and added studies of 2 families of their own in which these two tumors occurred simultaneously in 5 patients. Subsequently, the association between pheochromocytoma and MTC has been well confirmed in the literature.

Incidence

It is difficult to determine how commonly pheochromocytomas occur in association with MTC since different reports have emphasized different aspects of the MEN. It has become apparent that the incidence of pheochromocytomas is much higher in MEN IIb (neural tumors and MTC) than in MEN IIa (hyperparathyroidism and MTC). In MEN IIb, the incidence of pheochromocytoma usually approaches 50% [303], whereas in MEN IIa, it ranges from approximately 5 to 50% [127, 254, 350].

Clinical Manifestations

There are several distinctive features of pheochromocytomas when they occur in association with MTC. In this circumstance, bilateral and multifocal pheochromocytomas are very common and approximate a prevalence of 70% [303, 488, 534]. If hyperparathyroidism occurs, the pheochromocytoma is even more likely (84%) to be bilateral [451]. This contrasts with a bilateral incidence of usually less than 10% for sporadic pheochromocytomas [232, 267] and an incidence ranging from 20 to 50% in familial pheochromocytomas not associated with MTC [174, 254, 303, 488, 534]. Conversely, MTC is bilateral in essentially all patients in whom it occurs with a pheochromocytoma. Pheochromocytoma is much more likely to occur in patients with familial rather than sporadic MTC. In fact, although apparently sporadic cases of pheochromocytoma and MTC have been reported, it is very likely that these cases represent new mutations of the MEN II gene [79, 306, 488]. The presence of a pheochromocytoma in a patient with MEN II makes the presence of hyperparathyroidism more likely [254, 340]. By contrast, the presence of pheochromocytomas and mucosal neuromas in association with MTC make the incidence of hyperparathyroidism less likely [303]. The association of pheochromocytoma and hyperparathyroidism without MTC is rare [297] and the ectopic production of parathyroid hormone by pheochromocytomas, although reported, is not well documented [496]. Familial pheochromocytomas also occur in a distinct syndrome associated with other neural and vascular tumors but not associated with MTC. Also, MTC can rarely occur with pheochromocytomas that are apparently not familial [79, 488].

When pheochromocytomas and MTC occur together in the same patient, the MTC is likely to be diagnosed first [254]. The thyroid tumor may antedate the pheochromocytoma by as much as 21 years [200]. Multiple instances have been reported where a pheochromocytoma did not become clinically manifest until years after surgical removal of an MTC;

when this occurs, the MTC may not be recognized as part of MEN; furthermore a second pheochromocytoma may become manifest after removal of the first [254]. This sequence of events results in a greater incidence of pheochromocytomas in older patients with MTC. Less often, the thyroid and adrenal tumors may be discovered contemporaneously and, in some instances, the pheochromocytoma may be diagnosed before the MTC [330, 340, 352]. If hyperparathyroidism also exists, it too is likely to antedate the pheochromocytoma.

The incidence of malignancy does not seem to be increased in pheochromocytomas which occur in association with MTC [488]. In only 1 of 8 patients in one series was the pheochromocytoma considered to be malignant since histological deposits of tumor in the liver were indistinguishable from those of the adrenal tumor [254]. However, it is difficult to make the diagnosis of malignancy in pheochromocytomas [460]. Pheochromocytomas in MEN have been noted in accessory adrenal glands [330, 480] and at extra-adrenal sites, but the incidence of extra-adrenal pheochromocytomas is apparently not increased [342]. These pheochromocytomas can contain abnormal concentrations of calcitonin [244, 385, 517] arising from the adrenal cells or from metastatic MTC [357].

Diagnosis

Several authors have called attention to some atypical features of pheochromocytomas when they occur with MTC. Although not agreed to by all [298], the biochemical and clinical manifestations of pheochromocytoma associated with MTC may be more subtle than with sporadic pheochromocytomas [205, 488]. In one series, only 37% of patients with both tumors had symptoms of pheochromocytoma [254]. Sustained hypertension is less common in patients with MTC who have pheochromocytomas than in patients with sporadic pheochromocytomas [488], and basal and provocative biochemical testing are more likely to produce false-negatives [205, 330, 488]. This may be related to the fact that some pheochromocytomas, especially the small ones more likely to be diagnosed in MEN, have such a fast turnover of catecholamines that urinary metabolites may be normal or only slightly elevated.

Because of these subtle biochemical and clinical manifestations, the diagnosis of this potentially fatal tumor must be pursued aggressively in the patient with MTC when there is a high index of suspicion. The presence of a pheochromocytoma must be either established or ruled out since if present it must be treated before the thyroid tumor. If not, the patient is exposed to the well-known dangers of anesthesia-induced adrenergic crisis. The use of urinary VMA excretion studies, although useful in some instances, may not always be a reliable index for adrenal medullary hyperfunctioning, particularly in patients with early disease [479, 488]. Since there is a fast turnover rate of catecholamines in patients with small pheochromocytomas, urinary VMAs may only be slightly elevated despite an increased amount of urinary-free catecholamines [480]. In addition, total urinary-free catecholamines may be normal in patients with pheochromocytomas. The most sensitive biochemical test in these patients may be the urinary epinephrine/norepinephrine ratio. In a large kindred with familial MTC, an elevated epinephrine/norepinephrine ratio in 24-h urinary specimens proved to be of great value in screening family members for early adrenal medullary disease since epinephrine was found to be the major catecholamine secretory product in these patients [221]. The higher ratio of epinephrine to norepinephrine production may account for the lower incidence of sustained hypertension in patients with MTC, but this could represent a reporting bias.

Even in the presence of minimal symptoms and normal basal and functional tests of catecholamine secretion, the diagnosis of pheochromocytoma should not be overlooked in the patient with MTC. The presence of findings associated with pheochromocytomas such as MMN may dictate invasive procedures for the diagnosis of pheochromocytomas, such as venography and arteriography and adrenal echogram. However, these procedures may produce paroxysms, which can be fatal, and this risk must be weighed against the potential benefits. Thus, if diagnosis is to be aggressively pursued, the administration of blocking agents is recommended. Noninvasive procedures such as echography and scintigraphic portrayal may also be useful in diagnosis [512]. It is interesting to note in this context that MTC which produces catecholamines has been reported in association with pheochromocytomas [51, 517].

Adrenal Medullary Hyperplasia

Just as CCH may be a predecessor of MTC, so may adrenal medullary hyperplasia be a predecessor of the pheochromocytomas seen with MTC. This observation was made by *Ljungberg* [333] and emphasized by *Carney* et al. [115]. Although cases of adrenal medullary hyperplasia had been previously reported in the literature, none of them had been observed in patients with MTC [70, 198, 372, 460, 516]. *DeLellis* et al. [192] studied the adrenal glands of 10 patients from a large kindred with familial MTC and described many of the features of adrenal medullary hyperplasia. There is an increase in the medullary volume of adrenal glands when compared to controls. The increase in medullary mass results from diffuse and/or multifocal proliferation of adrenal medullary cells, primarily those found within the head and body of the glands. The multifocal proliferation can give an adenomatous appearance [333]. There is hypertrophy as well as hyperplasia of the cells and they show increased mitotic activity and increased total catecholamine content [115]. In addition, the ratio of epinephrine to norepinephrine is increased in the tumor [192].

These findings suggest that a sequence of events similar to those postulated for MTC and hyperparathyroidism, discussed later, take place in the development of pheochromocytomas – hypertrophy develops into hyperplasia and multifocal hyperplasia develops into nodularity. Neoplastic transformation to pheochromocytoma is the final stage of the sequence in most tumors, but malignant behavior can also be seen.

CCH can also develop as a secondary phenomenon in the patient with a chronic hypercalcemic state such as primary hyperparathyroidism [402]. This is thought to be due to stimulation of the C cells by the hypercalcemia. Under such circumstances, the C cells, stimulated at first, may eventually become exhausted, particularly in females who are calcitonin-deficient relative to males [402].

Hyperparathyroidism

History

There is an increased incidence of hyperparathyroidism in patients with MTC. It is of historical interest to note that in *Sipple's* [476] original description in 1961 of the association between thyroid carcinoma and pheochromocytoma, one of the patients had a parathyroid adenoma. However, *Sipple* [476] described this thyroid tumor as a follicular adenocarcinoma and did not comment on the distinct features of the MTC which had been described only 2 years earlier by *Hazard* et al. [242]. It was *Cushman* [152] who, in 1962, reported the familial occurrence of hyperparathyroidism (a parathyroid adenoma) in a patient with MTC who also had pheochromocytoma. The endocrine triad of MTC, bilateral pheochromocytomas and hyperparathyroidism, was more explicitly described by *Manning* et al. [340] in 1963. The case reported by *DeGraeff* et al. [189] in 1959 may represent the first reported association between these three tumors, but the thyroid cancer was diagnosed as being MTC in retrospective reviews by

Williams [536] and *Schimke and Hartmann* [453]. In the years following *Cushman's* [152] publication, additional reports of the association of hyperparathyroidism and MTC occurred [488]. *Steiner* et al. [488] added 2 additional cases in 1968 and brought the literature total to 13. In the intervening years, still more reports have clearly established the association.

Incidence

It is difficult to establish the exact incidence of hyperparathyroidism in patients with MTC. *Melvin* et al. [352] made the diagnosis of hyperparathyroidism in 10 of 12 patients of a kindred with MTC, whereas *Hill* et al. [254] could establish the diagnosis in only 2 of 73 patients with MTC. Additional series record an intermediate incidence [529] and the aggregate literature suggests an incidence or prevalence of 50% [122, 279, 529].

There are several possible explanations for this disparity. Reporting bias is undoubtedly a factor. Many studies are retrospective and occurred at a time when the association of MTC and hyperparathyroidism was not appreciated. Furthermore, there was, and still is, disagreement regarding the criteria necessary to distinguish between a normal and an abnormal parathyroid gland as well as the criteria to classify parathyroid abnormalities such as hypercellularity, hyperplasia, adenomatous hyperplasia, adenomas, and multiple adenomas. Another explanation for the variably reported incidence of hyperparathyroidism in MTC is the developing appreciation of the likelihood that MTC is associated with two distinct syndromes, MEN IIa and IIb, and that hyperparathyroidism is more common in MEN IIa and is less common, and perhaps even absent, in MEN IIb [303, 479]. Additionally, although the occurrence of hyperparathyroidism in the patient with MTC makes a pheochromocytoma more likely, parathyroid adenomas can occur in familial MTC in the absence of pheochromocytomas [337, 341]. As future reports take this distinction between MEN IIa and IIb into account, the association between MTC and hyperparathyroidism may become better quantified. Presently, the concurrence of hyperparathyroidism and MTC is well established but it must be considered as qualitative rather than quantitative. Even so, the presence of one tumor should always make the presence of the other suspect.

Pathology

The pathological findings in the parathyroid glands in patients with MTC are variable. When *Steiner* et al. [488] reviewed the literature in 1968, they recorded 10 cases with parathyroid adenoma and 3 cases with parathyroid hyperplasia, although two of the (multiple) adenomas may, in fact, have been hyperplasia. In later reports, the incidence of parathyroid hyperplasia in patients with MTC has increased and has even approached 100% [298, 350, 351]. This controversy regarding parathyroid pathology in patients with MTC is only reflective of the general difficulties in this area. There are no standardized histological criteria to distinguish the hyperfunctioning parathyroid gland from normal. There is a lack of uniform criteria to distinguish between parathyroid hyperplasia and parathyroid adenoma. This difficulty in evaluating the function of the parathyroid glands by histological examination is magnified by the evolving circumstances during which the parathyroid glands have been examined. In early studies, the parathyroid gland was examined in patients who had well-established clinical hyperparathyroidism. By contrast, because of the recent advances made in the diagnosis of hyperparathyroidism, many patients are now being brought to surgery relatively early in the course of their illness. This is especially so in the patient with MTC [350]. The use of the calcitonin assay in the diagnosis of MTC has resulted in the early surgical treatment of these patients. Since the thyroid changes can be so subtle in such patients, it is not surprising that the parathyroid changes can also be subtle. In fact, some of the histological changes of the parathyroid gland are too subtle to be categorized as hyperplasia; parathyroid glands thus involved have been called hypercellular [81].

Thus, parathyroid pathology in patients with MTC ranges from the most subtle changes of hypercellularity, through the changes of hyperplasia, to the relatively more distinct changes of adenoma. Multiple adenomas have been reported as well as the combination of adenoma and hyperplasia [81, 488]. This sequential pathology is reminiscent of the changes seen in the thyroid gland that vary from subtle histological changes, through CCH, to C-cell carcinoma, and in the changes in the adrenal medulla which include hyperplasia and pheochromocytoma and perhaps frank malignancy. The subtle changes of parathyroid hypercellularity may precede the development of hyperplasia and the hyperplasia may develop at the same or a different rate in each parathyroid gland. Parathyroid hyperplasia may be the final stage of the disorder in some cases, but hyperplasia may proceed (perhaps via adenomatous or nodular hyperplasia) to adenoma [10], similarly to the sequence of events postulated for the development of pheochromocytomas [192]. Although frank malignancy might represent the final step in this sequence, as it does in the thyroid gland with MTC and in rare cases of adrenal and neural tissue, parathyroid cancer has not been reported in MTC. It is likely, therefore, that the different histological stages of parathyroid hyperfunction that have been described in patients with MTC represent a natural history of parathyroid-gland pathology which corresponds to the natural history of the pathology of the thyroid and adrenal glands and perhaps even the neural tissue.

Functional Activity of the Parathyroid Gland in Patients with MTC

There may be no obvious relationship between the histological abnormality of the parathyroid gland, its secretory function, or clinical hyperparathyroidism. The presence of parathyroid hyperfunction as assessed by histological studies and measurements of parathyroid hormone concentrations in plasma does not correlate well with biochemical evidence of hyperparathyroidism such as hypercalcemia and hypophosphatemia or X ray or biochemical evidence of increased bone resorption such as osteopenia and increased plasma alkaline phosphatase [81, 298, 351]. For example, histological hyperparathyroidism may be associated with normocalcemia and normal plasma parathyroid hormone levels as well as hypercalcemia and elevated plasma parathyroid hormone levels [81, 298]. Hypercalcemia has been reported in patients with MTC who have normal parathyroid glands [81]. In such patients, it is possible that increased prostaglandins contributed to the hypercalcemia [274]. This lack of an obvious biological effect of the abnormal levels of parathyroid hormone may be partially explained by the equally abnormal concentrations of its physiological antagonist, calcitonin, in these patients. The ultimate biological effect of these two hormones may depend on their interactions. To fully evaluate their peripheral effect, it may be necessary to understand the status of the receptors of these two hormones after chronic exposure to hormone excess.

The manifestations of hyperparathyroidism may also depend on the nature of the secretory abnormality of the parathyroid gland. There is evidence to indicate that the functional status of the gland may change during the course of the disease. Part of this evidence comes from the histological studies of the parathyroid glands in these patients [81]. These histological studies suggest that there could be a progression of parathyroid pathology from hypercellularity to hyperplasia and including adenoma. Several clinical studies provide evidence that this histological progression, if it exists, may be accompanied by a functional progression of the parathyroid gland. *Keiser* et al. [298] observed that hypercalcemia seemed to be more common in patients with hyperplastic parathyroid glands than in patients with hypercellular glands. If so, the earliest changes in the histology of the parathyroid glands may not be accompanied by any biochemical changes in the patient. This may be influenced by the presence of calcitonin or by the secretory status of the parathyroid glands. In reference to the latter, *Heath* et al. [245] reported that in their group of patients with MEN IIa, hyperplasia of the parathyroid glands was accompanied by a lack of suppressibility by calcium infusion of plasma parathyroid hormone. This occurred despite the fact that basal calcium and parathyroid hormone were indistinguishable from controls. Perhaps lack of suppressibility of secretion despite normal baseline levels may be an early functional abnormality of the parathyroid gland and this subtle dysfunction is reflected by equally subtle biological effects of the hormone. This subtle abnormality may eventually progress to a frank abnormality in hormone secretion and the resulting clinical and biochemical changes. This view is supported by the findings of *Chong* et al. [127] who observed that clinical hypercalcemia and nephrolithiasis were more commonly seen in older patients with MEN. *Block* et al. [81] also noted that hypercalcemia is more likely to occur in the patients with adenoma than in the patients with hyperplasia.

Relationship to MTC

The link between hyperparathyroidism and MTC is more obscure than the embryological link between MTC and the other associated features, such as pheochromocytomas and neuromas. Two theories prevail. One possibility is that the hyperparathyroidism is a functional disorder which represents a compensatory response of the parathyroid glands to a hypocalcemic effect of calcitonin. This is supported by the finding of eucalcemia in most patients with MTC and the absence of radiological findings of hyperparathyroidism [298]. This view does have some additional clinical and experimental support. In patients with Paget's disease who are treated with pharmacological doses of calcitonin, secondary hyperparathyroidism has been described as a complicating effect of the calcitonin. However, hyperparathyroidism is not reported in all patients with Paget's disease who are treated with calcitonin [94]. In addition to the hyperparathyroidism mediated through calcitonin-induced hypocalcemia, there is some evidence to suggest that abnormal concentrations of calcitonin may have a direct stimulatory effect on parathyroid hormone secretion. This phenomenon has been reported both in vitro and in vivo [167].

There are data at least as convincing, however, that suggest that the hyperparathyroidism is an inherited rather than functional component of MTC. The high incidence of hyperparathyroidism in patients with MTC is just as compatible with this view as with the functional hypothesis. In some patients with MTC, hyperparathyroidism precedes the development of hypercalcitoninemia; by contrast, hyperparathyroidism may occur decades after the development of MTC [200]. In the patients with mucosal neuroma, there is a relatively low incidence of hyperparathyroidism despite the MTC [533]. This suggests a different genetic complex in these patients, because the absence of hyperparathyroidism is not compatible with a functional relationship between the thyroid and parathyroid tumors. The high incidence of hyperplasia, which is the common form of parathyroid abnormality in familial hyperparathyroidism (another genetically determined disorder), in patients with MTC also supports this view as does the higher incidence of hyperparathyroidism in patients with MTC who have the genetically associated bilateral pheochromocytomas.

The genetic view would be more attractive if the embryological origin of the parathyroid gland were neural crest as is the embryological origin of the other prominent features of the syndrome – pheochromocytoma, MTC, and MMN. Most evidence suggests that the parathyroid glands are of endodermal origin, arising from the third and fourth branchial pouches. However, there are recent data suggesting that the parathyroid glands may, indeed, be of neural-crest origin. If this is confirmed, it would provide a more unifying genetic basis for MEN.

A possible functional relationship between the parathyroid gland and the pheochromocytomas does not seem likely to explain the incidence of hyperparathyroidism in these patients. Although catecholamines can acutely stimulate parathyroid hormone secretion, and pheochromocytomas may ectopically secrete parathyroid hormone, patients with only pheochromocytomas do not have abnormal parathyroid function.

Mucosal Neuroma Syndrome

History

In 1966, *Williams and Pollock* [538] described 2 patients with MTC and pheochromocytomas who had neuromas involving the mucous membranes of the lips, tongue, and eyes; 9 similar cases were also discovered with a review of the literature [87, 219, 284, 335, 411, 424, 513]. Although they pointed out that the mucosal lesions were neuromas and not neurofibromas, they still incorrectly suggested that the MMN, pheochromocytomas, and MTC triad were closely related to von Recklinghausen's neurofibromatosis. *Ruppert* et al. [445] made the same suggestion in a similar case report. In 1967, *Ljungberg* et al. [330] observed a similar association in 14 patients and noted that 1 of them had a café-au-lait spot, later to be appreciated as an uncommon finding in this syndrome. In 1968, *Schimke* et al. [454] described 3 additional patients with this syndrome and recorded the presence of megacolon in each patient. In 1 patient, rectal biopsy was consistent with ganglioneuromatosis of the submucous and mysenteric plexuses. In the same year, *Gorlin* et al. [227] emphasized the association between MMN, pheochromocytomas, and MTC by reviewing 17 published cases of MMN. This review included the 11 cases of *Williams and Pollock* [538] and collected 6 additional cases from the literature. Approximately half of these patients had either pheochromocytomas, MTC, or both. In several patients the neuromas were either congenital or noticed within the first few years of life, thus becoming manifest before the other features of the syndrome. These authors also commented on the presence of a Marfanoid habitus, intestinal ganglioneuromatosis, diverticulosis, and medullated corneal nerve fibers in this group of patients. Thus, they articulated the features of the MMN, as currently accepted – MTC, pheochromocytoma, diffuse mucosal neuromas, and a Marfanoid habitus [228, 455]. MMN

Table II. Summary of clinical features in 41 patients with MEN III [from ref. 303]

Clinical features	Number of patients with findings			Number of patients with inadequate information
	positive	probable	negative	
Family history	14	2	15	10
Neuroma	41			
Oral	37		4	
Ocular	24		16	1
Others	4		36	1
'Bumpy' lips	35	2		4
Pheochromocytoma	19	4	18	
Unilateral	7			
Bilateral	12			
Medullary thyroid carcinoma	38		2	1
Marfanoid habitus	26	5		10
Hypertrophied corneal nerves	23			18
Skeletal defects	24		4	13
Gastrointestinal tract abnormalities	23		10	8

syndrome is also referred to as MEN IIb. *Baum and Adler* [51] emphasized the ophthalmological complications of this disorder which, in addition to the ocular neuromas and thickened corneal nerves, can include decreased tear function and an ectopic lacrimal punctum.

Several reasons have emerged for distinguishing MEN IIa from IIb [298, 303, 488]: (a) MMN and a Marfanoid habitus are not commonly found in MEN IIa; (b) hyperparathyroidism is not commonly found in MEN IIa; (c) MMN is not found in MTC-pheochromocytoma families who do not have MMN; and (d) in MEN IIb, family members do not have MTC or pheochromocytomas without having MMN. The review of *Khairi* et al. [303] of 41 patients with MMN showed 20 with pheochromocytoma and MTC, 18 with MTC, 3 with pheochromocytomas, 26 of 31 with Marfanoid features, 23 of 33 with gastrointestinal involvement, and only 1 with hyperparathyroidism, in contrast to the 57% incidence of the latter in MEN IIa (table II).

Pathology

Neuromas are the most consistent component of this syndrome. They have been reported to occur at most mucous membrane sites. The histological appearance of the neuromas has been best studied for the oral lesions, but studies of neuromas in other locations suggest a common pathological basis for all of the neural abnormalities. The most prominent microscopic feature of the neuromas is an increase in the size and number of nerves. Both medullated and unmedullated fibers are involved [229, 368]. The nerves are tortuous and highly branched and often surrounded by a thickened perineurium. This appearance is similar to that of a plexiform neuroma or an 'amputation' or 'traumatic' neuroma. Ganglion cells and connective tissue may be present but the latter is often not prominent. This usually distinguishes these neuromas from the neurofibromas of von Recklinghausen's neurofibromatosis which characteristically contain more connective tissue. However, the distinction between neuromas and neurofibromas cannot always be made [254, 302] and, in many cases of the MMN syndrome, cutaneous neurofibromas [214, 396, 445, 454] and even café-au-lait spots have been described [45, 233, 330, 376, 445]. Nevertheless, even the presence of neurofibromas would not establish a relationship between MMN and von Recklinghausen's neurofibromatosis since neurofibromas can also occur in other diseases not apparently related to MTC, such as von Hippel-Landau disease [265] and tuberous sclerosis [455]. Furthermore, MTC has not been described in either neurofibromatosis or von Hippel-Lindau disease despite the common feature of pheochromocytoma

[148, 265]. This association might be anticipated in future clinical studies. It is of interest to note that abnormal concentrations of calcitonin have been reported in neuromas [91, 517]. This finding is consistent with the common embryological origin of the cell types involved in MEN.

Mucosal Neuromas
Oral Cavity
Oral mucosal neuromas are most commonly found in a centrofacial distribution on the lips and tongue and on the buccal mucosa. This characteristic location can give the face of the patient an appearance common to all patients with the syndrome even though they are unrelated and of the opposite sex. In addition, this characteristic facial appearance along with the subsequently described Marfanoid habitus distinguish the patients from their normal siblings. The oral lesions are usually the first components of the syndrome to appear. They are almost invariably present by the first decade and can even be present at birth [230].

Lips. Usually both lips are affected but the upper one often appears more prominently involved, with central accentuation. The neuromas give the lips a prominent appearance which has been described as 'bumpy', 'nodular', 'blubbery', 'patulous', or 'papillomatous'. The inner aspects of the lips can also be involved.

Tongue. The centrofacial distribution of neuromas is also manifest in the tongue lesions which usually involve the tip and anterior third of the tongue dorsally (fig. 3). They can also extend along the lingual margin and in some cases have ventral or frenular location. The tongue neuromas range from pinhead size to several millimeters in diameter. They may become confluent and produce larger masses. The rare occurrences of macroglossia [366] or quadrangular [284] or bifid [424] abnormality of the tongue have been reported.

Buccal. These lesions are usually located posterior to each labial commissure. They can also be symmetrically arranged in relation to the premolars and on the pillars of the fauces [124].

Ocular Abnormalities
Mucosal neuromas can be present in the eyelids, conjunctiva, and cornea (fig. 4). The tarsal neuromas result in thickened eyelids and retracted eyelashes which give the eye a hooded and sleepy look. In addition to neuromas, a variety of other ocular abnormalities have been reported. The most common of these represents the manifestation of nerve proliferation

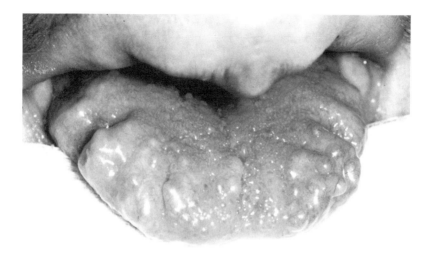

Fig. 3. Mucosal neuromas of tongue in a patient with MMN. Several large nodules are present on anterior third of tongue. Lateral margin exhibits a coarse undulating (left) and bumpy (right) appearance. Intraoral conical projections at angles of mouth are just visible bilaterally. Upper lip is diffusely thickened and exhibits characteristic central accentuation of thickening (from [118]).

in the cornea [51]. The medullated corneal nerves can be thickened and traverse the cornea and anastomose in the pupillary area [228, 303]. These hypertrophied nerve fibers are readily seen with the slit lamp but may occasionally be evident even on fundoscopic examination [305]. Although thickened corneal nerves may occur in other disorders, they occur in the majority of patients with MMN syndrome and are, therefore, an important clinical sign [434].

Other ocular abnormalities include stromal infiltrates, nodular perilimbal injections, impaired pupillary dilation, decreased tear formation with an abnormal Schimmer's test, and limbic neuromas [18, 50, 51, 455]. Despite the subsequently described Marfanoid habitus in these patients, lenticular subluxation has not been reported. The prominence of ocular abnormalities in this disorder provides important diagnostic findings.

Cutaneous Lesions

Neuromas. Cutaneous neuromas are seldom seen in MMN, in contrast to the common occurrence of cutaneous neurofibromas in von Reckling-

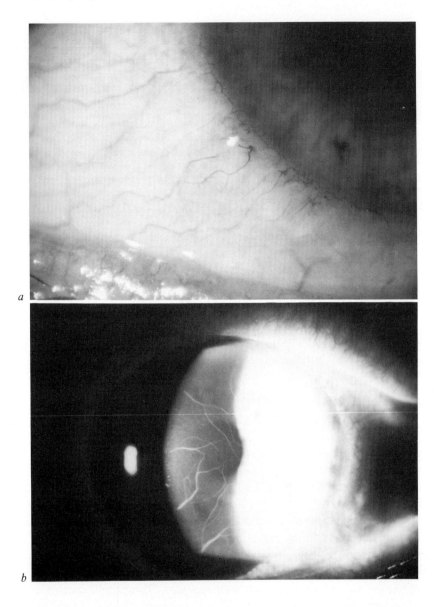

Fig. 4. a Plexiform neuroma of the bulbar conjuctiva at 7 o'clock near limbus, left eye from a patient with MMN (MEN II). *b* Highly visible corneal nerves, right eye [from ref. 51].

hausen's neurofibromatosis. However, both cutaneous neuromas [330, 505] and neurofibromas [330] have been reported in MMN; when they occur they are pedunculated, pigmented, and generally distributed [454].

Café-au-lait Spots. These are usually associated with von Recklinghausen's neurofibromatosis but have been observed in MMN [45, 233, 330, 376, 445, 454].

Gastrointestinal Abnormalities

Gastrointestinal abnormalities are prominent features of MMN [303]. Diarrhea is one of the most common symptoms of affected patients [254]. Its etiology is multifactorial and it was discussed in greater detail previously. Some of the gastrointestinal symptoms of diarrhea and constipation can additionally be ascribed to those gastrointestinal abnormalities which are part of the MMN syndrome. The most common of these is gastrointestinal ganglioneuromatosis [116, 119, 360].

The histological lesions of gastrointestinal ganglioneuromatosis are reminiscent of those that occur in the facial mucosa neuromas [303]. In fact, all cases of diffuse gastrointestinal ganglioneuromatosis occur in association with mucosal neuromas [533] and isolated intestinal ganglioneuromatosis is not associated with MTC [18]. Although some rectal biopsies are normal [454] or aganglionic [457, 504], there is usually a proliferation of the neural elements of the mysenteric (Auerbach's) and submucosal (Meissner's) plexuses. An eosinophilic infiltrate has been described in some reports [533]. The ganglioneuromatosis is best studied in the small and large intestine, but has also been observed in the esophagus and stomach [472, 533]. The anatomical lesions can be associated with the functional problems of swallowing difficulties, megacolon, diarrhea, and constipation [227, 533]. It is likely that the neurological lesions play some role in these functional gastrointestinal abnormalities [228, 455]. Symptoms and signs of gastrointestinal dysfunction seem to be relatively common in cases of ganglioneuromatosis. The functional consequence of the ganglioneuromatosis has not been clearly defined since both diarrhea and constipation commonly occur. However, it seems more likely that ganglioneuromatosis results in hypoactive intestinal function since constipation is more common [199] and the diarrhea seen in these patients can often be ascribed to one of the many humors produced by MTC. Additionally, megacolon is commonly reported in patients with intestinal ganglioneuromatosis, and diverticulosis can appear at an early age [454, 533]. Thus, this disorder should be considered in the differential diagnosis of Hirschsprung's disease [379].

Other Neural Abnormalities

Nerve proliferation has been observed in the bladder [445, 454, 542] and megaloureter has been reported along with megacolon [504]. This combination is explained by the common innervation received by the bladder and colon from the second to fourth sacral nerves [495]. Hyperplasia of the ganglion cells has also been noted in the bladder [445, 454]. Neuromas have also been reported in the bronchial tree [454, 542] and spinal nerve roots [229, 445, 454]. Enlargement of vagal and recurrent laryngeal nerves has been observed [45]. A variety of CNS tumors, including glioblastoma multiforme, gliomas, meningiomas, astrocytomas, and intracranial hemangiomas, have been reported in association with MMN as well as MEN [237, 302, 450].

Marfanoid Habitus

A Marfanoid habitus is commonly seen in the MMN syndrome; in one large series an incidence of 83% was observed [303]. The Marfanoid habitus refers to a tall, slender body with an abnormal upper-to-lower body segment ratio and poor muscular development (fig. 5). The extremities are thin and long, and there may be lax joints and hypotonic muscles [237]. Associated with the Marfanoid habitus may be dorsal kyphosis, pectus excavatum, pectus carinatum, pes cavus, and a high arched palate [18]. Pes cavus is uncommon in the true Marfan's syndrome whereas pes planus, common in Marfan's, is unreported in the Marfanoid habitus of MMN. Occurring less commonly are scoliosis [207] and lordosis [113], valgus deformities of the knees or toes [242], and abnormal metacarpal index [46, 364, 488]. In contrast to true Marfan's no patients with MMN have been reported to have aortic abnormalities, ectopia lentis, homocystinuria, or mucopolysaccharide abnormalities [227, 269, 303].

Other Musculoskeletal Abnormalities

Other associated musculoskeletal findings include aseptic necrosis of the lumbar spine [280, 454], slipped femoral epiphyses [91], dislocated hip [113, 333], ulnar deviation of the hands and wrists [91], inguinal hernia [228], abnormally shaped skull [228, 352], Scheuermann's disease [91], multiple osseous defects, and myopathy [114, 151]. The myopathy may be associated with either normal [303] or abnormal EEGs [324]. In 1 well-studied patient with MMN [151], the myopathy involved primarily the proximal musculature. Muscle biopsy showed scattered foci of myopathic degeneration; EM and histochemical studies revealed absence of normal

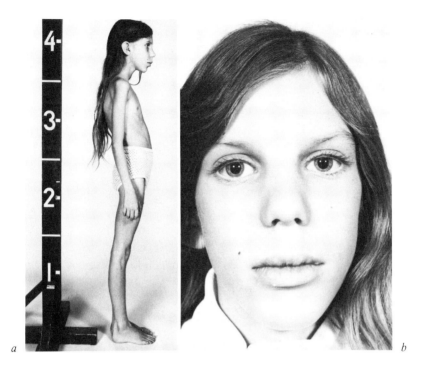

Fig. 5. a Marfanoid habitus, mild kyphosis, and poor muscle bulk, in a patient with MMN. *b* Close-up view of face of patient illustrating the characteristic facies of a patient with MMN [from ref. 91].

differentiation of muscle fibers. However, muscle biopsy can also be normal in patients with clinical myopathy [308]. Other less specific deformities such as osteopenia [91] and hypotonia [210] have also been described, and delayed puberty [280, 324, 454] has been reported in a few cases.

Other Locations of Neuromas and
Other Abnormalities of the Oral Cavity

Neuromas can occur on the mucosa of the palate, gingiva, nose, larynx, and pharynx [538].

The palate can be highly arched and narrow [354]. The teeth can also be abnormally shaped and have an extra cusp [28, 124]. These dental features along with neuromas can produce malocclusion [118]. Another explanation for the malocclusion is the prognathism which has been described for some patients [454]. However, the prognathism is due to soft tissue, not jawbone prominence.

Clinical Implications

The presence of facial neuromas and their resultant characteristic appearance are very important clinical features of the syndrome. Their importance derives from the fact that they are often the first clinical manifestations of the MMN syndrome since they can appear early in life and even be congenital. Because the neuromas can become clinically manifest up to 10 years before other features of MMN, they are important early clues to the presence of the other associated abnormalities [520]. Since at least two of the potentially lethal features of MEN – MTC and pheochromocytoma – are curable if discovered early, mucosal neuromas can be clinical signs which can lead to effective and even curative therapy in these patients.

By contrast, the neuromas themselves seldom require therapy. They do not undergo the malignant degeneration which can be seen with the neurofibromas, especially visceral, of von Recklinghausen's disease. In fact, the growth of neuromas seems to be self-limited by the second decade [118]. The occasional patient will seek surgical removal of the neuromas for cosmetic purposes and may, in fact, present to the physician in this manner. These cosmetic indications should not be underestimated. Many patients can be deeply disturbed by the appearance resulting from the neuromas.

Radiological Findings

Wallace et al. [521] and *Pearson* et al. [412] have summarized the prominent radiological findings in patients with MTC and its associated features. One of the most consistent findings is the presence of calcification in the neck tumor. This can be found in up to one-third of patients with MTC. Calcification of the tumor is found even more commonly in pathological studies but not all the calcified lesions are visible by X ray. The variable radiological appearance of the thyroidal calcifications reflects the variable size of the calcium deposits. This is distinguished from the more finely granular appearance of the psammomatous calcification seen in other thyroid cancers, especially the papillary variety.

Calcification can also be present in metastases of the MTC. This is especially common in the metastases to the regional nodes, but it can occur at almost any metastatic site. However, lung metastases do not seem to calcify. They instead produce a characteristic radiological picture of interstitial nodularity, prominent fibrosis and pulmonary hyperaeration. Pulmonary nodules are most prominent in the mid-lung fields. Other organs of the chest can be involved by the tumor since mediastinal metastases are relatively common. The tumor can invade the esophagus, the heart, and the great vessels of the thorax.

Bone lesions are commonly lytic, so the destructive properties of the tumor usually outweigh any inhibition of bone resorption that its high calcitonin concentration may produce. Bone lesions which are both lytic and blastic can also be seen. The bone changes of hyperparathyroidism are usually not seen by X ray even when this disease is present. However, two other radiological findings which are features of hyperparathyroidism, chondrocalcinosis and nephrolithiasis, can be present.

Angiographic studies of the tumor and its metastases reveal a high degree of vascularity. MTC does not concentrate radioiodine and the tumor appears as a 'cold spot' during thyroid scanning. Ultrasound studies reveal the commonly discrete nature of MTC.

The associated features of MTC also have their radiological manifestations. Pheochromocytomas are commonly bilateral in the patient with MTC and can be demonstrated by several procedures. The dangers of invasive procedures in a patient with pheochromocytoma should be kept in mind when considering these procedures in the patient with MTC. The pheochromocytomas in these patients can have the curvilinear 'egg shell' calcification characteristic of this tumor. However, suprarenal calcification of any form in the patient with MTC should alert the physician to the possible presence of a pheochromocytoma.

The gastrointestinal correlates of MTC can be documented by X ray. If the common complication of diarrhea is present, transit time for a barium meal is rapid. Increased transit time may also be present in the absence of diarrhea. Diarrhea can be contributed to by humoral as well as anatomical factors such as intestinal neurogangliomas in the patient with MTC. When intestinal ganglioneuromatosis is present, the intestinal wall is thickened because of the presence of these lesions. The large intestine may also be involved with diverticulosis. In the patient with constipation, megacolon may be present. Megaloureter has also been reported as has abnormal esophageal motility. Radiographic evidence of ulcer disease seems to be prominent in several reports of patients with MTC. Further studies are necessary to determine if this association is clinically significant [412, 521].

Treatment of Multiple Endocrine Neoplasia

The three major neoplastic disorders in MEN are MTC, pheochromocytoma, and hyperparathyroidism. All are potentially lethal, especially the first two, but may be cured if early diagnosis is made. Thus, aggressive treatment should be undertaken. Tumor markers, such as serial measurements of calcitonin, should be used to monitor the course of therapy [101, 117].

Medullary Thyroid Carcinoma

The treatment of choice for MTC is surgery [82]. Surgery must be designed with respect to each individual patient. Because of this, optimal treatment involves the participation of an experienced neck surgeon since many decisions regarding the procedure must be made during surgery [318]. With this consideration, the following generalities can be made. In tumor confined to the thyroid gland, total thyroidectomy should be performed. This is also true for the young patient [339]. This procedure is advocated even if there is apparent involvement of only one lobe since involvement of the other lobe is likely. This is especially true in familial MTC and MEN IIb where tumor growth can be aggressive [388]. Furthermore, bilaterality of tumor occurs often enough in seemingly sporadic cases to warrant total thyroidectomy. When even a rim of apparently normal thyroid tissue is left behind, the malignancy is likely to recur or appear at a metastatic site [80, 254, 298].

Even when there is regional lymph involvement, surgical treatment should be aggressive. Since the tumor can behave in an indolent manner, removal of regional lymphatics may retard the complications of locally invasive tumor. Therefore, the surgeon should be prepared to remove as many lymph nodes as possible. It may even be justified to remove an isolated metastatic lesion because of the indolent behavior of some tumors [254]. Therefore, a central dissection of lymph nodes is indicated [429] and neck dissection should be performed if the lateral cervical nodes are involved. The neck dissection may be a radical [429] or a modified procedure [298]. The sternum should be split for more extensive node removal if indicated by findings during surgery.

The importance of early and aggressive surgery is magnified by the difficulty encountered with reoperation. *Block* et al. [83] reported that calcitonin levels remained elevated in 6 of 8 patients following reexploration. Thus, second or subsequent procedures are not likely to remove remaining tumor, and even with an indolent course the prognosis worsens [83, 489].

Although ^{131}I, external radiation, and suppressive doses of T4 have been reported to be useful in some patients [196, 248, 519], these forms of treatment cannot be generally recommended on either clinical or theoretical grounds. The C cells do not concentrate iodine as do the follicular cells of the thyroid. The secretory activity of the tumor does not respond to either T4, TSH, or TRH [224]. However, most patients are ultimately placed on thyroid replacement after thyroidectomy. If normal thyroid is distributed in the tumor, it is remotely possible that a significant dose of radiation can be delivered to tumor cells if ^{131}I is given and concentrated by the remaining normal follicular cells. Such a procedure should only be considered as an adjunct to surgery in selected patients [161, 248].

Since surgery is beneficial in a majority of treatable patients, chemotherapy has not been extensively evaluated for MTC. With the increasing application of the calcitonin assay, early diagnosis can be made and under these circumstances surgical therapy is effective and even curative. It is, therefore, not likely that the need for chemotherapy as primary treatment will increase. Chemotherapy must be considered, however, in patients with metastatic disease and recurrent disease after optimal surgical treatment. Adriamycin has been used for such purposes. In one report of 5 patients treated with the drug, there was partial remission of tumor in 3 patients and in 2 patients diarrhea (presumably related to the presence of the tumor) disappeared [230]. Lack of success with adriamycin has also been reported [268] and chemotherapy has been generally disappointing in the MTC patient. However, newer agents and treatment regimens are still being evaluated [215, 462].

Pheochromocytoma

Every patient with MTC should be carefully evaluated for the presence of pheochromocytomas. Since pheochromocytomas associated with MTC may have subtle clinical and biochemical manifestations, as discussed earlier, the search for the tumor may have to be aggressive. If a pheochromocytoma is present, it shold be treated first in order to avoid any additional risk to the patient which would present during neck surgery. In addition, if metastases of the MTC are found during surgery for the pheochromocytoma, this may influence the final decision regarding surgery on the thyroid tumor. Consideration should be given to removal of both adrenal glands even if one appears normal because of the high incidence of bilaterality of the pheochromocytomas [328]. Even if the contralateral adrenal does not contain a pheochromocytoma at the time of surgery, one may ultimately develop and a paroxysm can be fatal.

Before surgery for the pheochromocytoma, the appropriate preoperative procedures should be instituted, such as fluid replacement and premedication with blocking agents. An abdominal approach will allow exploration for bilateral tumors and for metastatic thyroid tumor [293]. This consideration may outweigh the advantage of the bilateral flank approach which allows the adrenal veins to be more easily ligated before manipulation of the tumor. Also, the intraoperative use of blocking agents can reduce the danger of catecholamine release during surgical manipulation of the adrenals.

Neuromas

Most neuromas do not require surgical therapy. Malignant transformation of the neuromas, unlike some neurofibromas, does not seem to occur. However, the physician should not underestimate the cosmetic impact of facial neuromas and many patients justifiably seek therapy to improve their appearance. In fact, the need for cosmetic surgery may bring the patient to the physician for the first time.

Hyperparathyroidism

Surgical management of the parathyroid glands in the patient with MTC is difficult [81]. The pathology of the parathyroid glands is variable and there is no clear functional relationship between the gross or microscopic appearance of the parathyroid glands and their function. To consider the extremes, the patient with apparently normal glands may have biochemical evidence of hyperparathyroidism such as elevated plasma parathyroid hormone and calcium, and the patient with anatomical evidence of parathyroid hyperfunction may have normal plasma parathyroid hormone and calcium [352]. Because of such variable and confusing findings regarding parathyroid function in the patient with MTC, surgical management must be tailored to the individual patient by an experienced neck surgeon who is supported by a pathologist expert in this area. Functional test of parathyroid hormone secretion may be of some value in the preoperative evaluation of the patient [245]. The preoperative surgical plan must consider all of the clinical and biochemical features of the particular patient. The intraoperative management must be guided by these findings as well as by the operative findings, and the postoperative management should reflect the anticipated pathophysiological consequences of the particular surgical procedure, such as hypoparathyroidism as well as hypothyroidism.

In many patients, the surgical treatment of the MTC may be too extensive to preserve any parathyroid function. In the patient with normal plasma parathyroid hormone, calcium, and phosphorus, and with normal-appearing parathyroid glands, surgery should be designed to preserve as much parathyroid function as normal. But this should not be done at the risk of leaving behind residual thyroid tumor [437]. In the patient with elevated plasma parathyroid hormone and calcium, parathyroid tissue which appears abnormal should be removed. This may be relatively simple if the patient has a well-defined, single adenoma; it may be more difficult to identify hyperplastic glands and most difficult to identify hypertrophic glands [81]. If parathyroid hyperplasia is found, it should be kept in mind that this disorder usually involves all parathyroid glands, both eutopic and ectopic [352]. If the patient has no preoperative evidence of hyperparathyroidism, the physician should be conservative in his interpretation of parathyroid pathology. Under such circumstances, it is probably not justified to remove parathyroid glands which are minimally or questionably abnormal. However, clearly abnormal parathyroid glands in any clinical setting should be removed. The development of postoperative hypocalcemia must be anticipated since this can be a life-threatening complication. This will occur invariably in the patient who has had all parathyroid tissue removed and may occur, if only transiently, when all parathyroid tissue has been manipulated, something more likely to happen with more extensive procedures, especially those in which all parathyroid glands were biopsied or otherwise manipulated. Postoperative hypoparathyroidism may be a permanent or transient complication of surgery [81]. If the patient remains hypercalcemic despite surgery that seems appropriate, the possibility of ectopic parathyroid tissue should be considered. Parathyroid transplantation procedures in the management of parathyroid hyperplasia in the patient with MTC are of potential value and are under current evaluation.

References

1 Aach, R.; Kissane, J.: Medullary carcinoma of the thyroid with hypocalcemia and diarrhea. Am. J. Med. *46:* 961–971 (1969).
2 Abe, K.; Adachi, I.; Miyakawa, S.; Tanaka, M.; Yamaguchi, K.; Tanaka, N.; Kameya, T.; Shimosato, T.: Production of calcitonin, adrenocorticotropic hormone, and β-menocyte-stimulating hormone in tumors derived from amino precursor uptake and decarboxylation cells. Cancer Res. *37:* 4190–4194 (1977).
3 Abdullahi, S.E.; Arrigoni-Martelli, E.; Gramm, E.; Franco, L.; Velo, G.P.: Effect of calcitonin in different inflammatory models. Agents Actions *7:* 533–538 (1977).
4 Adachi, I.; Abe, K.; Tanaka, M.; Miyakawa, S.; Kumaoka, S.: Phosphaturic effect of i.v.-administered calcitonin in man. Endocr. jap. *21:* 317–322 (1974).
5 Adachi, I.; Abe, K.; Tanaka, M.; Yamaguchi, K.; Miyakawa, S.; Hirakawa, H.; Tanaka, N.: Plasma human calcitonin (hCT) levels in normal and pathologic conditions, and their responses to short calcium or tetragastrin infusion. Endocr. jap. *23:* 517–526 (1976).
6 Agar, J.W.; Rossetti, R.; Ethier, M.; Hull, J.D.; Pletka, P.G.: Interrelationship between calcitonin and prostaglandins A and E in medullary carcinoma of the thyroid. Clin. Res. *26:* 409A (1978).
7 Albores-Saavedra, J.; Rose, G.C.; Ibanez, M.L.; Russell, W.O.; Gray, C.E.; Dmochowski, L.: The amyloid in solid carcinoma of the thyroid gland. Lab. Invest. *13:* 77–89 (1964).
8 Albores-Saavedra, J.; Duran, M.E.: Association of thyroid carcinoma and chemodectoma. Am. J. Surg. *116:* 887–890 (1968).
9 Aldred, J.P.; Luna, P.D.; Zeedyk, R.A.; Bastian, J.W.: Inhibition by salmon calcitonin (sCT) of desoxycorticosterone acetate (DOCA)-induced hypertension in the rat. Proc. Soc. exp. Biol. Med. *152:* 557–559 (1976).
10 Aliapoulios, M.A.; Voelkel, E.F.; Munson, P.L.: Assay of human thyroid glands for thyrocalcitonin activity. J. clin. Endocr. Metab. *26:* 897–901 (1966).
11 Allannic, H.; Lorcy, Y.; Cornec, A.; Le Marec, B.; Calmettes, C.: Thyroid cancer with amyloid stroma. Results of a familial study. Annls Endocr. *41:* 34–41 (1980).
12 Alumets, J.; Hakanson, R.; Lundqvist, G.; Sundler, F.; Thorell, J.: Ontogeny and ultrastructure of somatostatin and calcitonin cells in the thyroid gland of the rat. Cell Tiss. Res. *206:* 193–202 (1980).
13 Amara, S.G.; Rosenfeld, M.G.; Birnbaum, R.S.; Roos, B.A.: Identification of the putative cell-free translocation product of rat calcitonin mRNA. J. biol. Chem. *255:* 2645–2648 (1980).
14 Anast, C.; David, L.; Winnacker, J.; Glass, R.; Baskin, W.; Brubaker, L.; Burns, T.: Serum calcitonin-lowering effect of magnesium in patients with medullary thyroid carcinoma. J. clin. Invest. *56:* 1615–1621 (1975).

15 Anderson, A.; Bergdahl, L.; Boquist, L.: Thyroid carcinoma in children. Am. Surg. *43:* 159–163 (1977).
16 Anderson, E.E.; Glenn, J.F.: Cushing's syndrome associated with anaplastic carcinoma of the thyroid gland. J. Urol. *95:* 1–4 (1966).
17 Anderson, R.J.; Wahner, W.H.: Symmetrical hypofunctioning thyroid nodules and medullary thyroid carcinoma. J. nucl. Med. *19:* 560 (1978).
18 Anderson, T.E.; Spackman, T.J.; Schwartz, S.S.: Roentgen findings in intestinal ganglioneuromatosis. Its association with medullary thyroid carcinoma and pheochromocytoma. Radiology *101:* 93–96 (1971).
19 Andersson, A.C.; Henningsson, S.; Jarhult, J.: Diamine oxidase activity and gamma-aminobutyric acid formation in medullary carcinoma of the thyroid. Agents Actions *10:* 299–301 (1980).
20 Ardaillou, R.; Fuagnat, P.; Milhaud, G.; Richet, G.: Effets de la thyrocalcitonine sur l'excrétion rénale des phosphates, du calcium et des ions H$^+$ chez l'homme. Nephron *4:* 298–314 (1967).
21 Ardaillou, R.; Fillastre, J.P.; Milhaud, G.; Rousselet, F.; Delaunay, F.; Richet, G.: Renal excretion of phosphate, calcium, and sodium during and after a prolonged thyrocalcitonin infusion in man. Proc. Soc. exp. Biol. Med. *131:* 56–60 (1969).
22 Ardaillou, R.; Sizonenko, P.; Meyrier, A.; Vallee, G.; Beaugas, C.: Metabolic clearance rate of radioiodinated human calcitonin in man. J. clin. Invest. *49:* 2345–2352 (1970).
23 Ardaillou, R.; Paillard, F.; Savier, C.; Bernier, A.: Renal uptake of radioiodinated human calcitonin in man. Revue eur. Etud. clin. biol. *16:* 1031–1036 (1971).
24 Ardaillou, R.: Kidney and calcitonin. Nephron *15:* 250–260 (1975).
25 Ardaillou, R.; Beaufils, M.; Nivez, M.P.; Mayaud, C.; Sraer, J.D.: Increased plasma calcitonin in early acute renal failure. Clin. Sci. mol. Med. *49:* 301–304 (1975).
26 Argemi, B.; Hours, M.C.; Kasbarian, M.; Grisoli, J.; Cannoni, M.; Codaccioni, J.L.; Simonin, R.: Localization of metastatic medullary thyroid carcinoma by immunoreactive calcitonin assay. Hormone metabol. Res. *9:* 248–249 (1977).
27 Arnalmonreal, F.M.; Goltzman, D.; Knaack, J.; Wang, N.S.; Huang, S.N.: Immunohistologic study of thyroidal medullary carcinoma and pancreatic insulinoma. Cancer *40:* 1060–1070 (1977).
28 Ascari, E.; Gamba, G.; Molinari, E.: La sindrome carcinoma midollare della tiroide, feocromocitoma a neuromi mucosi multipli. Contributo casistico. Minerva med., Roma *64:* 4418–4429 (1973).
29 Atkins, F.L.; Beaven, M.A.; Keiser, H.R.: Dopa decarboxylase in medullary carcinoma of the thyroid. New Engl. J. Med. *289:* 545–548 (1973).
30 August, G.P.; Shapiro, J.; Hung, W.: Calcitonin therapy of children with osteogenesis imperfecta. J. pediat. *91:* 1001–1005 (1977).
31 Augustin, R.; Hackeng, W.H.L.: Acute rise in serum calcitonin concentration during haemodialysis. Dt. med. Wschr. *103:* 508–512 (1978).
32 Aurbach, G.D.; Heath, D.A.: Parathyroid hormone and calcitonin regulation of renal function. Kidney int. *6:* 331–345 (1974).
33 Austin, L.A.; Heath, H., III: Calcitonin – physiology and pathophysiology. New Engl. J. Med. *304:* 269–278 (1981).
34 Austin, L.A.; Heath, H., III; Go, V.L.W.: Regulation of calcitonin secretion in normal

man by changes of serum calcium within the physiologic range. J. clin. Invest. *64:* 1721–1724 (1979).
35 Avioli, L.V.; Birge, S.J.; Scott, S.; Shieber, W.: Role of the thyroid gland during glucagon-induced hypocalcemia in the dog. Am. J. Physiol. *216:* 936–945 (1969).
36 Avioli, L.V.; Shieber, W.; Kipnis, D.M.: Role of glucagon and adrenergic receptors in thyrocalcitonin release in the dog. Endocrinology *88:* 1337–1340 (1971).
37 Avioli, L.V.: Vitamin D, the kidney, and calcium homeostasis. Kidney int. *2:* 241–246 (1972).
38 Avramides, A.; Baker, R.K.; Wallach, S.: Metabolic effects of synthetic salmon calcitonin in Paget's disease of bone. Metabolism *23:* 1037–1046 (1974).
39 Aw, T.C.; Seah, H.C.; Cheah, J.S.: Medullary carcinoma of the thyroid presenting as fatal asphyxia. Med. J. Aust. *i:* 188–189 (1979).
40 Baber, E.C.: Contributions to the minute anatomy of the thyroid gland of the dog. Phil. Trans. R. Soc. *166:* 557–568 (1876).
41 Baker, R.K.; Wallach, S.; Tashjian, A.H., Jr.: Plasma calcitonin in pycnodysostosis. Intermittently high basal levels and exaggerated responses to calcium and glucagon infusions. J. clin. Endocr. Metab. *37:* 46–55 (1973).
42 Baran, D.T.; Whyte, M.P.; Haussler, M.R.; Deftos, L.J.; Slatopolsky, E.; Avioli, L.V.: Effect of the menstrual cycle on calcium regulating hormones in the normal young woman. J. clin. Endocr. Metab. *50:* 437–439 (1980).
43 Barlet, J.P.; Garel, J.M.: Effect of an intravenous injection of cholecystokinin-pancreozymin on plasma calcium and calcitonin levels in newborn lambs. J. Endocr. *70:* 151–152 (1976).
44 Barrowman, J.A.; Bennett, A.; Hillenbrand, P.; Rolles, K.; Pollock, D.J.; Wright, J.T.: Diarrhoea in thyroid medullary carcinoma. Role of prostaglandins and therapeutic effect of nutmeg. Br. med. J. *iii:* 11–12 (1975).
45 Bartlett, R.C.; Myall, R.W.T.; Bean, L.R.; Mandelstam, P.: A neuropolyendocrine syndrome. Mucosal neuromas, pheochromocytomas, and medullary thyroid carcinoma. Oral Surg. oral Med. oral Path. *31:* 206–220 (1971).
46 Bartley, P.C.; Lloyd, H.M.; Aitken, R.E.: Marfanoid habitus and other abnormalities (Sipple's syndrome). Med. J. Aust. *ii:* 172–176 (1976).
47 Bates, R.F.; Bruce, J.B.; Care, A.D.: The effect of catecholamines on calcitonin secretion in the pig. J. Endocr. *46:* 11p (1970).
48 Bates, R.F.; Phillippo, M.; Lawrence, C.B.: The effect of propranolol on calcitonin secretion. J. Endocr. *48:* 8p (1970).
49 Baud, C.A.; deSiebenthal, J.; Langer, B.; Tupling, M.R.; Mach, R.S.: Effets de l'administration prolongée de thyrocalcitonine dans l'ostéoporose sénile humaine. Etude clinique, histologique et microradiographique. Schweiz. med. Wschr. *99:* 657–661 (1969).
50 Baum, J.L.: Abnormal intradermal histamine reaction in the syndrome of pheochromocytoma, medullary carcinoma of the thyroid gland, and multiple mucosal neuromas. New Engl. J. Med. *284:* 963–964 (1971).
51 Baum, J.L.; Adler, M.E.: Pheochromocytoma, medullary thyroid carcinoma, multiple mucosal neuroma. Archs ophthal., Chicago *87:* 574–584 (1972).
52 Baylin, S.B.; Beaven, M.A.; Engelman, K.; Sjoerdsma, A.: Elevated histamine activity in medullary carcinoma of the thyroid gland. New Engl. J. Med. *283:* 1238–1244 (1970).

53 Baylin, S.B.; Beaven, M.A.; Buja, L.M.; Keiser, H.R.: Histaminase activity. A biochemical marker for medullary carcinoma of the thyroid. Am. J. Med. *53:* 723–733 (1972).
54 Baylin, S.B.; Beaven, M.A.; Keiser, H.R.; Tashjian, A.H., Jr.; Melvin, K.E.W.: Serum histaminase and calcitonin levels in medullary carcinoma of the thyroid. Lancet *i:* 455–458 (1972).
55 Baylin, S.B.; Mendelsohn, G.; Weisburger, W.R.; Gann, D.S.; Eggleston, J.C.: Levels of histaminase and *l*-DOPA decarboxylase activity in the transition from C-cell hyperplasia to familial medullary thyroid carcinoma. Cancer *44:* 1315–1321 (1979).
56 Baylin, S.B.: Medullary carcinoma of the thyroid gland. Use of biochemical parameters in detection and surgical management of the tumor. Surg. Clins N. Am. *54:* 309–323 (1974).
57 Baylin, S.B.: Ectopic production of hormones and other proteins by tumors. Hosp. Pract. *10:* 117–126 (1975).
58 Baylin, S.B.; Gann, D.S.; Hsu, S.H.: Clonal origin of inherited medullary thyroid carcinoma and pheochromocytoma. Science *193:* 321–323 (1976).
59 Baylin, S.B.: Histaminase (diamine oxidase) activity in human tumors. An expression of a mature genome. Proc. natn. Acad. Sci. USA *74:* 883–887 (1977).
60 Baylin, S.B.; Bailey, A.L.; Hsu, S.H.; Foster, G.V.: Degradation of human calcitonin in human plasma. Metabolism *26:* 1345–1354 (1977).
61 Baylin, S.B.; Hsu, S.H.; Stevens, S.A.; Kallman, C.H.; Trump, D.L.; Beaven, M.A.: The effects of *l*-dopa on in vitro and in vivo calcitonin release from medullary thyroid carcinoma. J. clin. Endocr. Metab. *48:* 408–414 (1979).
62 Beaugie, J.M.; Belchetz, P.E.; Brown, C.L.; Frankel, R.J.; Lloyd, M.H.: Report of a family with inherited medullary carcinoma of the thyroid and pheochromocytoma. Br. J. Surg. *62:* 264–268 (1975).
63 Beaugie, J.M.; Brown, C.L.; Doniach, I.; Richardson, J.E.: Primary malignant tumours of the thyroid. The relationship between histological classification and clinical behaviour. Br. J. Surg. *63:* 173–181 (1976).
64 Becker, K.L.; Monaghan, K.G.; Silva, O.L.: Immunocytochemical localization of calcitonin in Kulchitsky cells of human lung. Archs Pathol. Lab. Med. *104:* 196–198 (1980).
65 Becker, K.L.; Silva, O.L.: Hypothesis. The bronchial Kulchitsky (K) cell as a source of humoral biologic activity. Med. Hypotheses *7:* 943 (1981).
66 Becker, K.L.; Snider, R.H.; Silva, O.L.; Moore, C.F.: Calcitonin heterogeneity in lung cancer and medullary thyroid carcinoma. Acta endocr., Copenh. *89:* 89–99 (1978).
67 Bell, N.H.: The effects of glucagon, dibutyryl cAMP and theophylline on calcitonin secretion in vitro. J. clin. Invest. *49:* 1368–1373 (1970).
68 Bernier, J.J.; Rambaud, J.C.; Cattan, D.; Prost, A.: Diarrhoea associated with medullary carcinoma of the thyroid. Gut *10:* 980–985 (1969).
69 Beskid, M.; Lorenc, R.; Rosciszewska: C-cell thyroid adenoma in man. J. Path. *103:* 1–4 (1971).
70 Bialestock, D.: Hyperplasia of the adrenal medulla in hypertension in children. Archs Dis. Childh. *36:* 465–473 (1961).
71 Bigazzi, M.R.; Revoltaella, S.; Casciano, S.; Vigneti, E.: High level of a nerve growth factor in the serum of a patient with medullary carcinoma of the thyroid gland. Clin. Endocrinol. *6:* 105–113 (1977).

72 Bijvoet, O.L.M.; Sluys Veer, J. van der; Vries, H.R. de; Koppen, A.T.J. van: Natriuretic effect of calcitonin in man. New Engl. J. Med. *284:* 681–688 (1971).
73 Birgenhager, J.C.; Upton, G.V.; Seldenrath, H.J.; Krieger, D.T.; Tashjian, A.H., Jr.: Medullary thyroid carcinoma. Ectopic production of peptides with ACTH-like, corticotrophin releasing factor-like, and prolactin production-stimulating activities. Acta endocr., Copenh. *83:* 280–292 (1976).
74 Blahos, J.: Calcitonin activity assessed by calcium tolerance test in patients with thyroid disorders. Endokrinologie *64:* 191–195 (1975).
75 Blahos, J.; Osten, J.; Mertl, L.; Kotas, J.; Gregor, O.; Reisenhauer, R.: Uricosuric effect of calcitonin. Hormone metabol. Res. *7:* 445–446 (1975).
76 Blahos, J.; Svoboda, Z.; Hoschel, G.: The effect of calcitonin on glucose metabolism. Endokrinologie *68:* 226–230 (1976).
77 Blaustein, A.: Calcitonin secreting struma-carcinoid tumor of the ovary. Human Pathol. *10:* 222–228 (1979).
78 Bloch-Michel, H.; Milhaud, G.; Coutris, G.; Waltzing, P.; Charret, A.; Morin, Y.; Verger, D.; Dussart, N.: Traitement au long cours de l'ostéoporose par la thyrocalcitonine. A propos de 7 observations. Revue Rhum. Mal. ostéo-artic. *37:* 629–638 (1970).
79 Block, M.A.; Horn, R.C., Jr.; Miller, J.M.; Barrett, J.L.; Brush, B.E.: Familial medullary carcinoma of the thyroid. Ann. Surg. *166:* 403–412 (1967).
80 Block, M.A.; Jackson, C.E.; Greenawald, K.A.; Yott, J.B.; Tashjian, A.H., Jr.: Clinical characteristics distinguishing hereditary from sporadic medullary thyroid carcinoma. Treatment implications. Archs Surg., Chicago *115:* 142–148 (1980).
81 Block, M.A.; Jackson, C.E.; Tashjian, A.H., Jr.: Management of parathyroid glands in surgery for medullary thyroid carcinoma. Archs Surg., Chicago *110:* 617–624 (1975).
82 Block, M.A.: Management of carcinoma of the thyroid. Ann. Surg. *185:* 133–144 (1977).
83 Block, M.A.; Jackson, C.E.; Tashjian, A.H., Jr.: Management of occult medullary thyroid carcinoma. Archs Surg., Chicago *113:* 368–372 (1978).
84 Bone, H.G.; Snyder, W.H.; McMillan, P.; Deftos, L.J.; Pak, C.Y.C.: Hyperabsorption of calcium in C-cell proliferative disorders. Proc. 60th Annu. Meeting Endocr. Soc. *A70:* 109 (1978).
85 Borle, A.B.: Regulation of cellular calcium metabolism and calcium transport by calcitonin. J. Membrane Biol. *21:* 125–146 (1975).
86 Braga, P.; Ferri, S.; Santagostino, A.; Olgiati, V.R.; Pecile, A.: Lack of opiate receptor involvement in centrally induced calcitonin analgesia. Life Sci. *22:* 971–978 (1978).
87 Braley, A.E.: Medullated corneal nerves and plexiform neuroma associated with pheochromocytoma. Trans. Am. ophthal. Soc. *52:* 189–196 (1954).
88 Brandenberg, W.: Metastasierender Amyloidkropf. Zentbl. allg. Path. path. Anat. *91:* 422–428 (1954).
89 Braunstein, H.; Stephens, C.L.: Parafollicular cells of human thyroid. Archs Path. *86:* 659–666 (1968).
90 Brewer, H.B., Jr.; Ronon, R.: Amino acid sequence of bovine thyrocalcitonin. Proc. natn. Acad. Sci. USA *63:* 940–947 (1969).
91 Brown, R.S.; Colle, E.; Tashjian, A.H., Jr.: The syndrome of multiple mucosal neuromas and medullary thyroid carcinoma in childhood. J. Pediat. *86:* 77–83 (1975).

92 Brown, W.H.: A case of pluriglandular syndrome. Lancet *ii:* 1022–1023 (1928).
93 Burk, W.: Über einen Amyloid-Tumor mit Metastasen; Dissertation (Pietzcker, Tübingen 1901).
94 Burkhardt, P.M.; Singer, F.R.; Potts, J.T., Jr.: Parathyroid function in Patients with Paget's disease treated with salmon calcitonin. Clin. Endocrinol. *2:* 15–22 (1973).
95 Bussolati, G.; Pearse, A.G.E.: Immunofluorescent localization of calcitonin in the C-cells of pig and dog thyroid. J. Endocr. *37:* 205–209 (1967).
96 Bussolati, G.; Foster, G.V.; Clark, M.B.; Pearse, A.G.E.: Immunofluorescent localization of calcitonin in medullary (C-cell) thyroid carcinoma using antibody to the pure porcine hormone. Virchows Arch. Abt. B Zellpath. *2:* 234–238 (1969).
97 Bussolati, G.; Monga, G.; Navone, R.; Gasparri, G.: Histochemical and electron-microscopical study of C-cells in organ culture; in Taylor, Calcitonin 1969. Proc. 2nd int. Symp., London 1969, pp. 240–251 (Heinemann, London 1970).
98 Bussolati, G.; Noorden, S.V.; Bordi, C.: Calcitonin- and ACTH-producing cells in a case of medullary carcinoma of the thyroid. Immunofluorescence investigations. Virchows Arch. Abt. A Path. Anat. *360:* 123–127 (1973).
99 Byfield, P.G.H.; McLoughlin, J.L.; Matthews, E.W.; MacIntyre, I.: A proposed structure for rat calcitonin. FEBS Lett. *65:* 242–245 (1976).
100 Calmettes, C.; Moukhtar, M.S.; Milhaud, G.: Correlation between calcitonin and carcinoembryonic antigen levels in medullary carcinoma of the thyroid. Biomed. Express *27:* 52–54 (1977).
101 Calmettes, C.; Moukhtar, M.S.; Milhaud, G.: Tumour markers. Calcitonin and carcinoembryonic antigen in medullary carcinoma of the thyroid. Presse méd. *8:* 3947–3950 (1979).
102 Calmettes, C.; Moukhtar, M.S.; Milhaud, G.: Plasma carcinoembryonic antigen versus plasma calcitonin in the diagnosis of medullary carcinoma of the thyroid. Cancer Immunol. Immunother. *4:* 251–256 (1978).
103 Canale, D.D.; Donabedian, R.K.: Hypercalcitoninemia in acute pancreatitis. J. clin. Endocr. Metab. *40:* 738–741 (1975).
104 Caniggia, A.; Gennari, C.; Bencini, M.; Cesari, L.; Borrello, G.: Calcium metabolism and 47-calcium kinetics before and after long-term thyrocalcitonin treatment in senile osteoporosis. Clin. Sci. *38:* 397–407 (1970).
105 Cantalamessa, L.; Catania, A.; Reschini, E.; Peracchi, M.: Inhibitory effect of calcitonin in growth hormone and insulin secretion in man. Metabolism *27:* 987–992 (1978).
106 Care, A.D.; Keynes, W.M.: The secretion of calcitonin by the parathyroid glands of the sheep. J. Endocr. *31:* xxxi–xxxii (1965).
107 Care, A.D.; Bates, R.F.L.; Gitelman, H.J.: A possible role for the adenyl cyclase system in calcitonin release. J. Endocr. *48:* 1–15 (1970).
108 Care, A.D.; Bates, R.F.L.; Swaminathan, R.; et al.: The role of gastrin as a calcitonin secretagogue. J. Endocr. *51:* 735–744 (1971).
109 Care, A.D.; Bell, N.H.; Bates, R.F.L.: The effect of hypermagnesemia on calcitonin secretion in vivo. J. Endocr. *51:* 381–386 (1971).
110 Care, A.D.; Bruce, J.B.; Boelkins, J.; Kenny, A.D.; Conaway, H.; Anast, C.S.: Role of pancreozymin, cholecystokinin and structurally related compounds as calcitonin secretagogues. Endocrinology *89:* 262 (1971).

111 Carey, D.E.; Jones, K.L.; Parthemore, J.G.; Deftos, L.J.: Studies of plasma calcitonin in agoitrous cretins. Clin. Res. *26:* 189A (1978).
112 Carey, D.E.; Jones, K.L.; Parthemore, J.G.; Deftos, L.J.: Calcitonin secretion in congenital thyroid dysgenesis. J. clin. Invest. *65:* 892–895 (1980).
113 Carman, C.T.; Brashear, R.E.: Pheochromocytomas as an inherited abnormality. Report of the tenth affected kindred and review of the literature. New Engl. J. Med. *263:* 419–423 (1960).
114 Carney, J.A.; Bianco, A.J., Jr.; Sizemore, G.W.; Hayles, A.B.: Multiple endocrine neoplasia with skeletal manifestations. J. Bone Jt Surg. *73A:* 405–410 (1981).
115 Carney, J.A.; Sizemore, G.W.; Tyce, G.M.: Bilateral adrenal medullary hyperplasia in MEN, type 2. Mayo Clin. Proc. *50:* 3–10 (1975).
116 Carney, J.A.; Go, V.L.; Sizemore, G.W.; Hayles, A.B.: Alimentary-tract ganglioneuromatosis. A major component of the syndrome of multiple endocrine neoplasia, type 2B. New Engl. J. Med. *295:* 1287–1291 (1976).
117 Carney, J.A.; Sizemore, G.W.; Hayles, A.B.: C-cell disease of the thyroid gland in multiple endocrine neoplasia, type 2b. Cancer *44:* 2173–2183 (1979).
118 Carney, J.A.; Sizemore, G.W.; Lovestedt, S.A.: Mucosal ganglioneuromatosis, medullary thyroid carcinoma, and pheochromocytoma: MEN, type 2B. Oral Surg. *41:* 739–752 (1976).
119 Carney, J.A.; Sizemore, G.W.; Hayles, A.B.: Multiple endocrine neoplasia, type 2B. Pathobiol. Annu. *8:* 105–153 (1978).
120 Carney, J.A.; Sizemore, G.W.; Sheps, S.G.: Adrenal medullary disease in multiple endocrine neoplasia, type 2. Am. J. Path. *66:* 279–290 (1979).
121 Castells, S.; Ju, C.; Baker, R.K.; Wallach, S.: Effects of synthetic salmon calcitonin osteogenesis imperfecta. Curr. ther. Res. *16:* 1–14 (1974).
122 Catalona, W.J.; Engelman, K.; Ketcham, A.S.; Hammond, W.G.: Familial medullary thyroid carcinoma, pheochromocytoma, and parathyroid hormone (Sipple's syndrome). Cancer *28:* 1245–1254 (1971).
123 Catherwood, B.D.; Deftos, L.J.: Presence by radioimmunoassay of a calcitonin-like substance in porcine pituitary glands. Endocrinology *106:* 1886–1891 (1980).
124 Cernea, P.; Crepy, C.; Kuffer, R.: Neuromes myéliniques muqueux de la cavité buccale. Revue Stomat. Chir. maxillofac. *68:* 103–116 (1967).
125 Chiba, S.; Himori, N.: Effects of salmon calcitonin on SA nodal pacemaker activity and contractility in isolated, blood-perfused atrial and papillary muscle preparations of dogs. Jap. Heart J. *18:* 214–220 (1977).
126 Chiba, T.; Taminato, T.; Kadowaki, S.; Goto, Y.; Mori, K.; Seino, Y.; Abe, H.; Chihara, K.; Matsukura, S.; Fujita, T.; Kondo, T.: Effects of (Asu1,7)-331 calcitonin on gastric somatostatin and gastrin release. Gut *21:* 94–97 (1980).
127 Chong, G.C.; Beahrs, O.H.; Sizemore, G.W.; Woolner, L.H.: Medullary carcinoma of the thyroid gland. Cancer *35:* 695–704 (1975).
128 Clark, M.B.; Byfield, P.G.H.; Boyd, G.W.; Foster, G.V.: A radioimmunoassay for human calcitonin M. Lancet *ii:* 74–77 (1969).
129 Clark, M.B.; Williams, C.C.; Nathanson, B.M.; Horton, R.E.; Glass, H.I.; Foster, G.V.: Metabolic fate of human calcitonin in the dog. J. Endocr. *61:* 199–210 (1974).
130 Clark, O.H.; Rehfeld, S.J.; Deftos, L.J.; Castner, B.; Stroop, J.; Loken, H.F.: Iodine

deficiency produces hypercalcemia and hypercalcitoninemia in rats. Surgery, St. Louis 83: 626–632 (1978).
131 Cochran, M.; Hillyard, C.J.; Dew, G.J.; Martin, T.J.: Acute responsiveness to calcitonin in chronic renal failure. Br. med. J. ii: 396–398 (1976).
132 Cohen, S.L.; MacIntyre; I.; Grahame-Smith, D.; Walker, J.G.: Alcohol-stimulated calcitonin release in medullary carcinoma of the thyroid. Lancet ii: 1172–1174 (1973).
133 Cohn, S.H.; Dombrowski, C.S.; Hauser, W.; Klopper, J.; Atkins, H.L.: Effects of porcine calcitonin on calcium metabolism in osteoporosis. J. clin. Endocr. Metab. 33: 719–728 (1971).
134 Condon, J.R.; Ives, D.; Knight, M.J.; Day, J.: The aetiology of hypocalcemia in acute pancreatitis. Br. J. Surg. 62: 115–118 (1975).
135 Coombes, R.C.; Greenberg, P.B.; Hillyard, C.; MacIntyre, I.: Plasma-immunoreactive-calcitonin in patients with non-thyroid tumors. Lancet i: 1080–1083 (1974).
136 Cooper, C.W.; Schwesinger, W.H.; Mahgoub, A.M.; Ontjes, D.A.: Thyrocalcitonin. Stimulation of secretion by pentagastrin. Science 172: 1238–1240 (1971).
137 Cooper, C.W.: Ability of several cations to promote secretion of thyrocalcitonin in the pig. Proc. Soc. exp. Biol. Med. 148: 449–454 (1975).
138 Cooper, C.W.; Obie, J.F.; Toverud, S.U.; Munson, P.L.: Elevated serum calcitonin and serum calcium during suckling in the baby rat. Endocrinology 101: 1657–1664 (1977).
139 Cooper, C.W.; Obie, J.F.: Hypocalcemia and increased serum calcitonin in baby rats given glucose orally. Proc. Soc. exp. Biol. Med. 157: 374 (1978).
140 Cooper, C.W.; Peng, T.-C.; Obie, J.F.; Garner, S.C.: Calcitonin-like immunoreactivity in rat and human pituitary glands. Histochemical, in vitro, and in vivo studies. Endocrinology 107: 98–107 (1980).
141 Copp, D.H.; Cameron, E.C.; Cheney, B.A.; Davidson, A.G.F.; Henze, K.G.: Evidence for calcitonin – a new hormone from the parathyroid that lowers blood calcium. Endocrinology 70: 638–649 (1962).
142 Copp, D.H.; Crockroft, D.W.; Kueh, Y.: Ultimobranchial origin of calcitonin. Hypocalcemic effect of extracts from chicken glands. Can. J. Physiol. Pharmacol. 45: 1095–1099 (1967).
143 Copp, D.H.; Crockroft, D.W.; Kueh, Y.: Calcitonin from ultimobranchial glands of dogfish and chickens. Science 158: 924–925 (1967).
144 Cramer, C.F.; Parkes, C.O.; Copp, D.H.: The effect of chicken and hog calcitonin on some parameters of Ca, P, and Mg metabolism in dogs. Can. J. Physiol. Pharmacol. 47: 181–184 (1969).
145 Cramer, C.F.: Effect of salmon calcitonin on in vivo calcium absorption in rats. Calcif. Tissue Res. 13: 169–172 (1973).
146 Cramer, S.F.; Bradshaw, R.A.; Baglan, N.C.; Meyers, J.A.: Nerve growth factor in medullary carcinoma of the thyroid. Human Pathol. 10: 731–736 (1979).
147 Croughs, R.J.M.; Eastham, W.N.; Hackeng, W.H.L.; Schopman, W.; Feltkamp-Vroom, T.M.; Dolman, A.; Hennemann, G.: ACTH and calcitonin secretion in medullary carcinoma of the thyroid. Clin. Endocrinol. 1: 157–171 (1972).
148 Crowe, B.W.; Schull, W.J.; Neel, J.V.: Multiple neurofibromatosis (Thomas, Springfield 1956).
149 Cummings, J.H.; Milton-Thompson, G.J.; Billings, J.; Newman, A.; Misiewicz, J.J.:

Studies on the site of production of diarrhea induced by prostaglandins. Clin. Sci. mol. Med. *46:* 15P (1974).

150 Cunliffe, W.J.; Black, M.M.; Hall, R.; Johnston, I.D.A.; Hudgson, P.; Shuster, S.; Gudmundsson, T.V.; Joplin, G.F.; Williams, E.D.; Woodhouse, N.J.Y.; Galante, L.; MacIntyre, I.: A calcitonin-secreting thyroid carcinoma. Lancet *ii:* 63–66 (1968).

151 Cunliffe, W.J.; Hudgson, P.; Fulthorpe, J.J.; Black, M.M.; Hall, R.; Johnston, I.E.A.; Shuster, S.: A calcitonin-secreting medullary thyroid carcinoma associated with mucosal neuromas, marfanoid features, myopathy, and pigmentation. Am. J. Med. *48:* 120–126 (1970).

152 Cushman, P., Jr.: Familial endocrine tumors. Report of two unrelated kindred affected with pheochromocytomas, one also with multiple thyroid carcinomas. Am. J. Med. *32:* 352–360 (1962).

153 Cutler, G.B., Jr.; Habener, J.F.; Dee, P.C.; Potts, J.T., Jr.: Radioimmunoassay for chicken calcitonin. FEBS Lett. *38:* 209–212 (1974).

154 Cutler, G.B., Jr.; Habener, J.F.; Potts, J.T., Jr.: Comparative immunochemical studies of chicken ultimobranchial calcitonin. Gen. compar. Endocr. *24:* 183–190 (1974).

155 Dangoumau, J.; Bussière, C.; Noël, M.; Balabaud, C.: Influence of calcitonin on bile production in the rat. J. Pharmacol. *7:* 69–76 (1976).

156 David, L.; Cohn, D.V.; Anast, C.S.: Identification of artifacts introduced to radioimmunoassays by the treatment of sera with charcoal in order to obtain 'dehormonized' sera. Applications to the radioimmunoassay of parathormone and calcitonin. Path. Biol., Paris *23:* 833–834 (1975).

157 David, L.; Salle, B.; Chpard, P.; Grafmeyer, D.: Studies on circulating immunoreactive calcitonin in low birth weight infants during first 48 hours of life. Helv. paediat. Acta *32:* 39–48 (1977).

158 DeCourcy, J.L.; DeCourcy, C.B.: Pheochromocytoma and the general practitioner, p. 90 (Newman, Cincinnati 1952).

159 Deftos, L.J.; Lee, M.R.; Potts, J.T., Jr.: A radioimmunoassay for thyrocalcitonin. Proc. natn. Acad. Sci. USA *60:* 293–299 (1968).

160 Deftos, L.J.: An immunoassay for human calcitonin. I. The method. Metabolism *20:* 1122–1128 (1971).

161 Deftos, L.J.; Stein, M.F.: Radioiodine as an adjunct to the surgical treatment of medullary thyroid carcinoma. J. clin. Endocr. Metab. *50:* 967–968 (1980).

162 Deftos, L.J.; Bury, A.E.; Habener, J.F.; Singer, F.R.; Potts, J.T., Jr.: Immunoassay for human calcitonin. II. Clinical studies. Metabolism *20:* 1129–1137 (1971).

163 Deftos, L.J.; Goodman, A.D.; Engelman, K.; Potts, J.T., Jr.: Suppression and stimulation of calcitonin secretion in medullary thyroid carcinoma. Metabolism *20:* 428–431 (1971).

164 Deftos, L.J.; Habener, J.F.; Mayer, G.P.; Bury, A.E.; Potts, J.T., Jr.: An immunoassay for bovine calcitonin. J. Lab. clin. Med. *79:* 480–490 (1972).

165 Deftos, L.J.; Powell, D.; Parthemore, J.G.; Potts, J.T., Jr.: Secretion of calcitonin in hypocalcemic states in man. J. clin. Invest. *52:* 3109–3114 (1973).

166 Deftos, L.J.: Radioimmunoassay for calcitonin in medullary thyroid carcinoma. J. Am. med. Ass. *227:* 403–406 (1974).

167 Deftos, L.J.; Parthemore, J.G.: Secretion of parathyroid hormone in patients with medullary thyroid carcinoma. J. clin. Invest. *54:* 416–420 (1974).

168 Deftos, L.J.; Watts, E.G.; Copp, D.H.; Potts, J.T., Jr.: A radioimmunoassay for salmon calcitonin. Endocrinology 94: 155–160 (1974).
169 Deftos, L.J.; Roos, B.A.; Bronzert, D.; Parthemore, J.G.: Immunochemical heterogeneity of calcitonin in plasma. J. clin. Endocr. Metab. 40: 407–410 (1975).
170 Deftos, L.J.; Ross, B.A.; Parthemore, J.G.: Calcium and skeletal metabolism. The Western Journal of Medicine 123: 447–458 (1975).
171 Deftos, L.J.: Plasma calcitonin measurement as a marker for cancer. West. J. Med. 124: 489–490 (1976).
172 Deftos, L.J.; Boorman, G.A.; Roos, B.A.: Immunoassay of calcitonin in rat medullary thyroid carcinoma. Hormone metab. Res. 8: 83–84 (1976).
173 Deftos, L.J.; McMillan, P.J.; Sartiano, G.P.; Abuid, J.; Robinson, A.G.: Simultaneous ectopic production of parathyroid hormone and calcitonin. Metabolism 25: 543–550 (1976).
174 Deftos, L.J.: Calcitonin in clinical medicine. Adv. internal Med. 23: 159–193 (1978).
175 Deftos, L.J.: Regulations of skeletal homeostasis. Clin. Res. 26: 528A (1978).
176 Deftos, L.J.: The thyroid gland in skeletal and calcium metabolism; in Avioli, Krane, Metabolic bone disease, pp. 447–487 (Academic Press, New York 1978).
177 Deftos, L.J.; Burton, D.; Bone, H.G.; Catherwood, B.D.; Parthemore, J.G.; Moor, R.Y.; Minick, S.; Guillemin, R.: Immunoreactive calcitonin is present in the intermediate lobe of the pituitary gland. Life Sci. 23: 743–748 (1978).
178 Deftos, L.J.; Burton, D.; Catherwood, B.D.; Bone, H.G.; Parthemore, J.G.; Guille, R.; Watkins, W.; Moore, R.Y.: Demonstration by immunoperoxidase histochemistry of calcitonin in the anterior lobe of the rat pituitary. J. clin. Endocr. Metab. 47: 457–460 (1978).
179 Deftos, L.J.; Catherwood, B.D.; Bone, H.G.; Parthemore, J.G.; Minick, S.; Guillemin, R.: Pituitary calcitonin. Clin. Res. 26: 629A (1978).
180 Deftos, L.J.; O'Riordan, J.L.H.: Problems in radioassays of the calcitropic hormones; in Copp, Talmage, Endocrinology of calcium metabolism. Proc. 6th Parathyroid Conf., pp. 345–348 (Excerpta Medica, Amsterdam 1978).
181 Deftos, L.J.; Krook, L.; Mayer, G.P.: Plasma calcitonin in the bovine species. Proc. Soc. exp. Biol. Med. 162: 150–156 (1979).
182 Deftos, L.J.; First, B.P.: Calcitonin as a drug. Ann. intern. Med. 95: 192–197 (1981).
183 Deftos, L.J.: Immunohistological studies of non-thyroidal calcitonin-producing tumors. J. clin. Endocr. Metab. 50: 1042–1045 (1980).
184 Deftos, L.J.; Parthemore, J.G.; Price, P.A.: Changes in plasma bone Gla protein during treatment of bone disease. Calcif. Tissue int. 34: 121–124 (1982).
185 Deftos, L.J.; Weisman, M.H.; Williams, G.H.; Karpf, D.B.; Frumar, A.M.; Davidson, B.H.; Parthemore, J.G.; Judd, H.L.: Age- and sex-related changes of calcitonin secretion in humans. New Engl. J. Med. 302: 1351–1351 (1980).
186 Deftos, L.J.; Burton, D.W.; Watkins, W.B.; Catherwood, B.D.: Immunohistological studies of artiodactyl and teleost pituitaries with antisera to calcitonin. Gen. compar. Endocr. 42: 9–18 (1980).
187 Deftos, L.J.; Bone, H.G.; Parthemore, J.G.: Immunohistological studies of medullary thyroid carcinoma and C-cell hyperplasia. J. clin. Endocr. Metab. 51: 857–862 (1980).

188 Deftos, L.J.; Catherwood, B.D.: Dissociation between ACTH and beta endorphin immunoreactivity in cells of the rat pituitary. Life Sci. *27:* 223–228 (1980).
189 DeGraeff, J.; Muller, H.; Moolenaar, A.: Phaeochromocytoma. A report of seven cases. Acta med. scand. *164:* 419 (1959).
190 DeLellis, R.A.: The formaldehyde induced fluorescence technique for the demonstration of biogenic amines in diagnostic histopathology. Cancer *28:* 1704–1710 (1971).
191 DeLellis, R.A.; Wolfe, J.H.: Calcitonin in spindle cell thymic carcinoid tumors. Archs Pathol. Lab. Med. *100:* 340 (1976).
192 DeLellis, R.A., Wolfe, H.J.; Gagel, R.F.; Feldman, Z.T.; Miller, H.H.; Gang, D.L.; Reichlin, S.: Adrenal medullary hyperplasia. Am. J. Path. *83:* 177–196 (1976).
193 DeLellis, R.A.; Nunnemacher, G.; Wolfe, H.J.: C-cell hyperplasia – an ultrastructural analysis. Lab. Invest. *36:* 237–248 (1977).
194 DeLellis, R.A.; Rule, A.H.; Spiler, I.; Nathanson, L.; Tashjian, A.H., Jr.; Wolfe, H.J.: Calcitonin and carcinoembryonic antigen as tumor markers in medullary thyroid carcinoma. Am. J. clin. Path. *70:* 587–594 (1978).
195 Dent, C.E.; Faccini, J.; Hodsman, A.: Medullary carcinoma of thyroid gland in a girl aged 10 years. Archs Dis. Childh. *51:* 223–226 (1976).
196 Didolker, M.S.; Moore, G.E.: Hormone-dependent medullary carcinoma of the thyroid. Am. J. Surg. *128:* 100–102 (1974).
197 Donahower, G.F.; Schumacher, O.P.; Hazard, J.B.: Medullary carcinoma of the thyroid – a cause of Cushing's syndrome. Report of two cases. J. clin. Endocr. *28:* 1199–1204 (1968).
198 Drukker, W.; Formijne, P.; Schoot, J.B.: Hyperplasia of the adrenal medulla. Br. med. J. *i:* 186–189 (1957).
199 Duffy, T.J.; Erickson, E.C.; Jordan, G.L.; Bennett, H.D.: Megacolon and bilateral pheochromocytoma. Am. J. Gastroent., N.Y. *38:* 555–563 (1962).
200 Dunn, E.L.; Nishiyama, R.H.; Thompson, N.W.: Medullary carcinoma of the thyroid gland. Surgery, St Louis *73:* 848–858 (1973).
201 Dymling, J.F.; Ljungberg, O.; Hillyard, C.J.; Greenberg, P.B.; Evans, I.M.A.; MacIntyre, I.: Whiskey. A new provocative test for calcitonin secretion. Acta endocr., Copenh. *82:* 500–509 (1976).
202 Dyson, B.C.: Cushing's disease. Report of a case associated with carcinoma of the thyroid gland and cryptococcosis. New Engl. J. Med. *261:* 169–172 (1959).
203 Economiddu, J.; Karacoulis, P.; Manousos, O.N.; Manesis, E.; Kydenakis, A.; Doutras, D.A.: Carcinoembryonic antigen in thyroid disease. J. clin. Path. *30:* 878–880 (1977).
204 Ellison, E.H.; Wilson, S.D.: The Zollinger-Ellison syndrome. Reappraisal and evaluation of 260 registered cases. Ann. Surg. *160:* 512–530 (1964).
205 Engelman, K.; Horwitz, D.; Ambrose, I.M.; Sjoerdsma, A.: Further evaluation of the tyramine test for pheochromocytoma. New Engl. J. Med. *278:* 705 (1968).
206 Ericson, L.E.: Degranulation of the parafollicular cells of the rat thyroid by vitamin D_2-induced hypercalcemia. J. Ultrastruct. Res. *24:* 145–149 (1968).
207 Falck, B.; Larson, B.; Mecklenburg, C.V.; Rosenberg, E.; Svenaeus, K.: On the presence of a second specific cell system in mammalian thyroid gland. Acta physiol. scand. *62:* 491–492 (1964).
208 Fawell, W.N.; Thompson, G.: Nutmeg for diarrhea of medullary carcinoma of thyroid. New Engl. J. Med. *289:* 108–109 (1973).

209 Finegold, M.J.; Haddad, J.R.: Multiple endocrine tumors. Report of an unusual triad in a family. Archs Path. 76: 449–455 (1963).
210 Flensborg, E.W.; Schiotz, P.O.: Abnormal intradermal histamine reaction and elevated serum calcitonin in the syndrome of: Medullary carcinoma of the thyroid gland, pheochromocytoma, and multiple mucosal neuromas. Dan. med. Bull. 21: 20–26 (1974).
211 Foster, G.W.; Baghdiantz, A.; Kumar, M.A.; Slack, E.; Soliman, H.A.; MacIntyre, I.: Thyroid origin of calcitonin. Nature, Lond. 202: 1303–1305 (1964).
212 Foster, G.V.; MacIntyre, I.; Pearse, A.G.E.: Calcitonin production and the mitochondrion-rich cells of the dog thyroid. Nature, Lond. 203: 1029–1030 (1964).
213 Foster, G.V.; Byfield, P.G.H.; Gudmundsson, T.V.: Calcitonin. Clin. Endocrinol. Metab. 1: 93–124 (1972).
214 Franssila, K.: Value of histologic classification of thyroid cancer. Acta pathol. microbiol. scand. 225: 66 (1971).
215 Freeman, A.I.; Ettinger, L.J.; Brecher, M.L.: cis-Dichlorodiammineplatinum (II) in childhood cancer. Cancer Treat. Rep. 63: 1615–1620 (1979).
216 Freed, W.J.; Perlow, M.J.; Wyatt, R.J.: Calcitonin. Inhibitory effect on eating in rats. Science 206: 850–851 (1979).
217 Fritsch, H.; Noorden, S. Van; Pearse, A.: Localization of somatostatin-substance P- and calcitonin-like immunoreactivity in the neural ganglion of Ciona intestinalis L. Cell Tiss. Res. 202: 262–274 (1979).
218 Fritsch, H.; Noorden, S. Van; Pearse, A.: Calcitonin-like immunochemical staining in the alimentary tract of Ciona intestinalis L. Cell Tiss. Res. 205: 439–444 (1980).
219 Frobese, C.: Das aus arK-haltigen Nervenfasern bestehende, ganglienzellenlose, echte Neurom in Rankenform. Zugleich ein Beitrag zu den nervösen Geschwülsten der Zunge und des Augenlides. Virchows Arch. path. Anat. Physiol. 240: 312–327 (1923).
220 Fujimoto, Y.; Oka, A.; Fukumitsu, M.; Obara, T.; Akisada, M.; Yamaguchi, K.: Physical and radiological findings specific for medullary carcinoma of the thyroid gland. Endocr. jap. 22: 225–232 (1975).
221 Gagel, F.R.; Melvin, K.E.W.; Tashjian, A.H., Jr.; Miller, H.H.; Feldman, Z.T.; Wolfe, H.J.; DeLellis, R.A.; Cervi-Skinner, S.; Reichlin, S.: Natural history of the familial medullary thyroid carcinoma-pheochromocytoma syndrome and the identification of preneoplastic stages by screening studies. A five-year report. Trans. Ass. Am. Physns 88: 177–191 (1975).
222 Galante, L.; Gudmundsson, T.V.; Matthews, E.W.; Williams, E.D.; Tse, A.; Woodhouse, N.J.Y.; MacIntyre, I.: Thymic and parathyroid origin of calcitonin in man. Lancet ii: 537–539 (1968).
223 Goldberg, W.M.; McNeil, M.J.: Cushing's syndrome due to an ACTH-producing carcinoma of the thyroid. Can. med. Ass. J. 96: 1577–1579 (1967).
224 Goltzman, D.; Potts, J.T., Jr.; Ridgeway, E.C.; Maloof, F.: Calcitonin as a tumor marker. Use of a radioimmunoassay for calcitonin in the postoperative evaluation of patients with medullary thyroid carcinoma. New Engl. J. Med. 290: 1035–1039 (1974).
225 Goltzman, D.; Tischler, A.S.: Characterization of the immunochemical forms of calcitonin released by a medullary thyroid carcinoma in tissue culture. J. clin. Invest. 61: 449–458 (1978).

226 Gorevic, P.D.; Rosenthal, C.J.; Franklin, E.C.: Amyloid-related serum component (SAA). Studies in acute infections, medullary thyroid carcinoma, and postsurgery. Clin. Immunol. Immunopathol. *6:* 83–93 (1976).
227 Gorlin, R.J.; Sedano, H.O.; Vickers, R.A.; Cervenka, J.: Multiple mucosal neuromas, pheochromocytomas, and medullary carcinoma of the thyroid – a syndrome. Cancer *22:* 293–299 (1968).
228 Gorlin, R.J.; Vickers, R.A.: Multiple mucosal neuromas, pheochromocytoma, medullary carcinoma of the thyroid, and Marfanoid body build with muscle wasting. Re-examination of a syndrome of neural crest malmigration. Birth Defects, Orig. Article Ser., vol. II, pp. 69–72 (1971).
229 Gorlin, R.J.; Mirkin, B.L.: Multiple mucosal neuromas, pheochromocytoma, medullary carcinoma of the thyroid, and Marfanoid body build with muscle wasting. Z. Kinderheilk. *113:* 313–325 (1972).
230 Gottlieb, J.A.; Hill, C.S., Jr.: Adriamycin (NSC-123127) therapy in thyroid carcinoma. Cancer Chemother. Rep. *6:* 283–296 (1975).
231 Gowenlock, A.H.; Platt, D.S.; Campbell, A.C.P.; Wormsley, K.G.: Oat-cell carcinoma of bronchus secreting 5-hydroxytryptophan. Lancet *i:* 304–306 (1964).
232 Graham, J.B.: Pheochromocytoma and hypertension. Analysis of 207 cases. Int. Abstr. Surg. *92:* 105 (1951).
233 Grandboise, J.; Tchou, P.K.: Mucous membrane manifestations of neurofibromatosis with malignant tumors of endocrine glands. Cutis *5:* 1235–1239 (1969).
234 Gray, T.K.; Bieberdorf, F.A.; Fordtran, J.S.: Thyrocalcitonin and the jejunal absorption of calcium, water, and electrolytes in normal subjects. J. clin. Invest. *52:* 3084–3088 (1973).
235 Gray, T.K.; Brannan, P.; Juan, D.; Morawski, G.; Fordtran, J.S.: Ion transport changes during calcitonin-induced intestinal secretion in man. Gastroenterology *71:* 392–398 (1976).
236 Graze, K.; Spiler, I.J.; Tashjian, A.H., Jr.; Melvin, K.E.W.; Cervi-Skinner, S.; Gagel, R.F.; Miller, H.H.; Wolfe, H.J.; DeLellis, R.A.; Leape, L.; Feldman, Z.T.; Reichlin, S.: Natural history of familial medullary thyroid carcinoma. New Engl. J. Med. *299:* 980–985 (1978).
237 Greenhouse, A.: Pheochromocytoma and meningioma of the foramen magnum. Ann. intern. Med. *55:* 124–127 (1961).
238 Grimley, P.M.; Deftos, L.J.; Weeks, J.R.; Rabson, A.S.: Growth in vitro and ultrastructure of cells from a medullary carcinoma of the human thyroid gland. Transformation by simian virus 40 and evidence of thyrocalcitonin and prostaglandin production. J. natn. Cancer Inst. *42:* 663–690 (1969).
239 Guttmann, S.; Pless, J.; Sandrin, E.; Jaquenoud, P.A.; Bossert, H.; Willems, H.: Synthèse des thyrocalcitonins. Helv. chim. Acta *51:* 1155–1161 (1968).
240 Hargis, G.K.; Williams, G.A.; Reynolds, W.A.; Chertow, B.S.; Kukreja, S.C.; Bowser, E.N.; Henderson, W.J.: Effect of somatostatin on parathyroid hormone and calcitonin secretion. Endocrinology *102:* 745–750 (1978).
241 Hayles, A.B.; Kennedy, R.L.; Beahrs, O.H.; Woolner, L.B.: Management of a child with thyroidal carcinoma. J. Am. med. Ass. *173:* 21–28 (1960).
242 Hazard, J.B.; Hawk, W.A.; Crile, G.: Medullary (solid) carcinoma of the thyroid – a clinicopathologic entity. J. clin. Endocr. Metab. *19:* 152–161 (1969).

243 Hazard, J.B.: The C-cells (parafollicular cells) of the thyroid gland and medullary thyroid carcinoma. Am. J. Path. *88:* 213–250 (1977).
244 Heath, H., III; Edis, A.J.: Pheochromocytoma associated with hypercalcemia and ectopic secretion of calcitonin. Ann. intern. Med. *91:* 208–210 (1979).
245 Heath, H.; Sizemore, G.W.; Carney, J.A.: Preoperative diagnosis of occult parathyroid hyperplasia by calcium infusion in patients with multiple endocrine neoplasia, type 2A. J. clin. Endocr. Metab. *43:* 428–435 (1976).
246 Heath, H.; Sizemore, G.W.: Plasma calcitonin in normal man. Differences between men and women. J. clin. Invest. *60:* 1135–1140 (1977).
247 Heersche, J.N.M.; Marcus, R.; Aurbach, G.D.. Calcitonin and the formation of 3′,5′-AMP in bone and kidney. Endocrinology *94:* 241–247 (1974).
248 Hellman, D.E.; Kartchner, M.; Antwerp, J.D. Van; Salmon, S.E.; Patton, D.D.; O'Mara, R.: Radioiodine in the treatment of medullary carcinoma of the thyroid. J. clin. Endocr. Metab. *48:* 451–455 (1979).
249 Hennessey, J.F.; Gray, T.K.; Cooper, C.W.; Ontjes, D.A.: Stimulation of thyrocalcitonin secretion by pentagastrin and calcium in two patients with medullary carcinoma of the thyroid. J. clin. Endocr. Metab. *36:* 200–203 (1973).
250 Hennessey, J.F.; Wells, S.A.; Ontjes, D.A.; Cooper, C.W.: A comparison of pentagastrin injection and calcium infusion as provocative agents for the detection of medullary thyroid carcinoma. J. clin. Endocr. Metab. *39:* 487–495 (1974).
251 Hesch, R.D.; Hufner, M.; Hasenjager, M.; Cruetzfeldt, W.: Inhibition of gastric secretion by calcitonin in man. Hormone metabol. Res. *3:* 140 (1971).
252 Heynen, G.; Kanis, J.A.; Oliver, D.; Ledingham, J.G.G.; Russell, R.G.G.: Evidence that endogenous calcitonin protects against renal bone disease. Lancet *ii:* 1322–1325 (1976).
253 Heynen, G.; Franchimont, P.: Human calcitonin radioimmunoassay in normal and pathological conditions. Eur. J. clin. Invest. *4:* 213–222 (1974).
254 Hill, C.S., Jr.; Ibanez, M.L.; Samaan, N.A.; Ahearn, M.J.; Clark, R.L.: Medullary (solid) carcinoma of the thyroid gland. An analysis of the M.D. Anderson Hospital experience with patients with the tumor, its special features, and its histogenesis. Medicine, Baltimore *52:* 141–171 (1973).
255 Hillyard, C.J.; Coombes, R.C.; Greenberg, P.B.; Galante, L.S.; MacIntyre, I.: Calcitonin in breast and lung cancer. Clin. Endocrinol. *5:* 1–8 (1976).
256 Hillyard, C.J.; Evans, I.; Hill, P.; Taylor, S.: Familial medullary thyroid carcinoma. Lancet *i:* 1009–1010 (1978).
257 Hirsch, P.F.; Gauthier, G.F.; Munson, P.L.: Thyroid hypocalcemic principle and recurrent laryngeal nerve injury as factors affecting the response to parathyroidectomy in rats. Endocrinology *73:* 244–252 (1963).
258 Hirsch, P.F.; Voelkel, E.F.; Munson, P.L.: Thyrocalcitonin. Hypocalcemic hypophosphatemic principle of the thyroid gland. Science *146:* 412–413 (1964).
259 Hirsch, P.F.; Munson, P.L.: Thyrocalcitonin. Physiol. Rev. *49:* 548–622 (1969).
260 Hokfelt, B.; Sjorgren, B.; Falkheden, T.: Steroid hormone production in a case of Cushing's syndrome with electrolyte changes simulating primary aldosterone. Acta endocr., Copenh. *31:* 175–184 (1959).
261 Hokfelt, T.; Efendic, S.; Hellerstrom, C.; Johansson, O.; Luft, R.; Arimura, A.: Cellular localization of somatostatin in endocrine-like cells and neurons of the rat with

special references to the A1-cells of the pancreatic islets and to the hypothalamus. Acta endocr., Copenh. *80:* 5–41 (1975).
262 Holtrop, M.E.; Raisz, L.G.; Simmons, H.A.: The effects of parathyroid hormone, colchicine and calcitonin on the ultrastructure and the activity of osteoclasts in organ culture. J. Cell Biol. *60:* 346–355 (1974).
263 Horn, R.C., Jr.: Carcinoma of the thyroid. Description of a distinctive morphological variant and report of seven cases. Cancer *4:* 697–707 (1951).
264 Horton, E.W.; Main, I.H.M.; Thompson, C.J.; Wright, P.M.: Effect of orally administered prostaglandin E on gastric secretion and gastrointestinal motility in man. Gut *9:* 655–658 (1968).
265 Horton, W.A.; Wong, V.; Eldridge, R.: Von Hippel-Lindau disease. Clinical and pathological manifestations in nine families with 50 affected members. Archs intern. Med. *136:* 769–777 (1976).
266 Hotz, J.; Goebell, H.: Similar modes of action of calcitonin and glucagon in inhibiting pancreatic enzyme secretion in man. Klin. Wschr. *57:* 1265–1272 (1979).
267 Hume, D.M.: Pheochromocytoma in the adult and in the child. Am. J. Surg. *99:* 458–496 (1960).
268 Husain, M.; Alsever, R.N.; Lock, J.P.; George, W.F.; Katz, F.H.: Failure of medullary carcinoma of the thyroid to respond to doxorubicin therapy. Hormone Res. *9:* 22–25 (1978).
269 Hutchins, G.M.; Mirvis, S.E.; Mendelsohn, G.; Bulkley, G.H.: Supravalvular aortic stenosis with parafollicular cell (C-cell) hyperplasia. Am. J. Med. *64:* 967–973 (1978).
270 Ibanez, M.L.; Russell, W.O.; Albores-Saavedra, J.; Lampertico, R.; White, E.C.; Clark, R.L.: Thyroid carcinoma-biologic behavior and mortality-postmortem findings in 42 cases, including 27 in which the disease was fatal. Cancer *19:* 1039–1052 (1966).
271 Ibanez, M.L.; Cole, V.W.; Russell, W.O.; Clark, R.I.: Solid carcinoma of the thyroid gland. Cancer *20:* 706–723 (1967).
272 Isaac, R.; Nivez, M.P.; Piamba, G.; Fillastre, J.P.; Ardaillou, R.: Influence of calcium infusion on calcitonin and parathyroid hormone concentrations in normal and hemodialyzed subjects. Clin. Nephrol *3:* 14–17 (1975).
273 Isaac, R.; Merceron, R.; Caillens, G.; Raymond, J.P.; Ardaillou, R.: Effects of calcitonin on basal and thyrotropin-releasing hormone-stimulated prolactin secretion in man. J. clin. Endocr. Metab. *50:* 1011–1015 (1980).
274 Isaacs, P.; Whittaker, S.M.; Turnberg, L.A.: Diarrhea associated with medullary carcinoma of the thyroid. Studies of intestinal function in a patient. Gastroenterology *67:* 521–526 (1974).
275 Ishikawa, N.; Hamada, S.: Association of medullary carcinoma of thyroid with carcinoembryonic antigen. Br. J. Cancer *34:* 111–115 (1976).
276 Israel, L.; Depierre, A.; Calmettes, C.; Milhaud, G.: The secretion of calcitonin by testicular carcinomas. Nouv. Presse méd. *6:* 3866–3870 (1977).
277 Ivey, J.L.; Roos, B.A.; Shen, F.H.; Baylink, D.J.: Increased immunoreactive CT in idiopathic hypercalciuria. Metab. Bone Dis. related Res. *3:* 39–42 (1981).
278 Jackson, C.E.; Block, M.A.; Greenawald, K.A.; Tashjian, A.H., Jr.: The two-mutational-event theory in medullary thyroid cancer. Am. J. hum. Genet. *31:* 704–710 (1979).

279 Jackson, C.E.; Tashjian, A.H., Jr.; Block, M.A.: Diagnostic dependability of calcitonin assay in family studies for medullary thyroid carcinoma. J. Lab. clin. Med. *78:* 817–818 (1971).
280 Jacobi, H.; Kleine–Natrop, H.E.: Beitrag zum Syndrom der angeborenen fibrillaren Neurome. Derm. Mschr. *156:* 644–652 (1970).
281 Jaquet, A.J.: Ein Fall von metastasierenden Amyloidtumoren (Lymphosarcoma). Virchows Arch. path. Anat. Physiol. *185:* 251–267 (1906).
282 Jensen, M.K.; Transbol, I.; Olesen, K.H.: Cancer of Cushing's syndrome. Nord. Med. *73:* 197–224 (1965).
283 Johnston, I.D.A.; Watson, A.J.: Surgical implications of medullary carcinoma of the thyroid. Proc. R. Soc. Med. *63:* 612–615 (1970).
284 Jorge, J.M.; Brachetto-Brian, D.: Contribution à l'étude des formes incomplètes de la maladie de Recklinghausen (considérations histologiques et pathogéniques). Bull. Ass. fr. Etude Cancer *16:* 158–170 (1927).
285 Kakudo, K.; Miyauchi, A.; Katayama, S.: Ultrastructural study of thyroid medullary carcinoma. Acta path. jap. *27:* 605–622 (1977).
286 Kakudo, K.; Miyauchi, A.; Ogihara, T.; Takai, S.I.; Kitamura, M.; Kosaki, G.; Kumahara, Y.: Medullary carcinoma of the thyroid. Giant cell type. Archs Pathol. Lab. Med. *102:* 445–447 (1978).
287 Kalina, M.; Foster, G.V.; Clark, M.B.; Pearse, A.G.E.: C-cells in man; in Taylor, Calcitonin 1969. Proc. 2nd int. Symp., London 1969, pp. 268–273 (Heinemann, London 1970).
288 Kalina, M.; Pearse, A.G.E.: Ultrastructural localization of biogenic amines in C-cells of mice; in Taylor, Calcitonin 1969. Proc. 2nd int. Symp., London 1969, pp. 262–267 (Heinemann, London 1970).
289 Kalu, D.N.; Hadji-Georgopoulos, A.; Foster, G.V.: Evidence for physiological importance of calcitonin in the regulation of plasma calcium in rats. J. clin. Invest. *55:* 722–727 (1975).
290 Kameya, T.; Shimosato, T.; Adachi, I.; Abe, K.; Kasai, N.; Kimura, K.; Baba, K.: Immunohistochemical and ultrastructural analysis of medullary carcinoma of the thyroid in relation to hormone production. Am. J. Path. *89:* 555–574 (1977).
291 Kanis, J.A.; Earnshaw, M.; Heynen, G.; Russell, R.G.G.; Woods, C.G.: The possible role of calcitonin deficiency in the development of bone disease due to chronic renal failure. Calcif. Tissue Res. *22:* 147–153 (1977).
292 Kaplan, E.L.; Peskin, G.W.; Arnaud, C.D.: Non-steroid, calcitonin-like factor from the adrenal gland. Surgery, St Louis *66:* 167–174 (1969).
293 Kaplan, E.L.; Peskin, G.W.: Physiologic implications of medullary carcinoma of the thyroid gland. Surg. Clins N. Am. *51:* 125–137 (1971).
294 Kaplan, E.L.; Sizemore, G.; Hill, B.J.; Peskin, G.W.: Calcitonin in nonthyroid tumors in man. Clin. Res. *20:* 724 (1972).
295 Kaplan, E.L.: Recent advances in calcium metabolism. Application to medullary carcinoma of the thyroid and hyperparathyroidism. Surg. A. *5:* 97–119 (1973).
296 Kedar, I.; Ravid, M.; Sohar, E.: In vitro synthesis of 'amyloid' fibrils from insulin, calcitonin, and parathormone. Israel J. med. Scis *12:* 1137–1140 (1976).
297 Keene, J.E.; Correa, R.J., Jr.: Pheochromocytoma associated with parathyroid adenoma. Report of a case and review of the literature. J. Urol. *106:* 443–447 (1971).

298 Keiser, J.R.; Beaven, M.A.; Doppman, J.; Wells, S.; Buja, L.M.: Medullary thyroid carcinoma, pheochromocytoma, and parathyroid disease. Ann. intern. Med. 78: 561–579 (1973).
299 Keusch, G.; Binswanger, U.; Dambacher, M.A.; Fischer, J.A.: Ectopic ACTH syndrome and medullary thyroid carcinoma. Acta endocr., Copenh. 86: 306–316 (1977).
300 Keutmann, H.T.; Parsons, J.A.; Potts, J.T., Jr.; Schlueter, R.J.: Isolation and chemical properties of two calcitonins from salmon ultimobranchial glands. J. biol. Chem. 245: 1491–1496 (1970).
301 Keutmann, H.T.; Lequin, R.M.; Habener, J.F.; Singer, F.R.; Niall, H.D.; Potts, J.T., Jr.: Chemistry and physiology of the calcitonins. Some recent advances; in Taylor, Endocrinology 1971. Proc. 3rd int. Symp., London 1972, pp. 316–323 (Heinemann, London 1972).
302 Keynes, W.M.; Till, A.S.: Medullary carcinoma of the thyroid gland. Q. Jl Med. 40. 443–456 (1971).
303 Khairi, M.R.A.; Dexter, R.N.; Burzynski, N.J.; Johnston, C.C., Jr.: Mucosal neuroma, pheochromocytoma, and medullary thyroid carcinoma. MEN, type III. Medicine, Baltimore 54: 89–112 (1975).
304 Kisloff, B.; Moore, E.W.: Effects of intravenous calcitonin on water, electrolyte, and calcium movement across in vivo rabbit jejunum and ileum. Gastroenterology 73: 462–468 (1977).
305 Knox, D.E.; Prout, T.E.: Enlarged corneal nerves in the syndrome of medullary carcinoma of the thyroid, multiple neuromas, and in some bilateral pheochromocytoma. Birth Defects 7: 161–163 (1971).
306 Knudson, A.G., Jr.; Strong, L.C.: Mutation and cancer. Neuroblastoma and pheochromocytoma. Am. J. hum. Genet. 24: 514–532 (1972).
307 Koelz, H.R.; Drack, G.T.; Blum, A.L.: Effect of calcitonin on salivary amylase secretion in vivo and in vitro. Schweiz. med. Wschr. 106: 298–299 (1976).
308 Koke, M.P.; Braley, A.E.: Bilateral plexiform neuromata of the conjunctiva and medullate corneal nerves. Am. J. Ophthal. 23: 179 (1940).
309 Korman, M.G.; Lauer, M.C.; Hansky, J.: Hypergastrinemia in chronic renal failure. Br. med. J. i: 209–210 (1972).
310 Kracht, J.; Hachmeister, U.; Christ, U.: C-cells in the human thyroid; in Taylor, Calcitonin 1969. Proc. 2nd int. Symp., London 1969, pp. 274–280 (Heinemann, London 1970).
311 Krane, S.M.; Harris, E.D., Jr.; Singer, F.R.; Potts, J.T., Jr.: Acute effects of calcitonin on bone formation in man. Metabolism 22: 51–58 (1973).
312 Krook, L.; Lutwak, L.; McEntee, K.: Dietary calcium ultimobranchial tumors and osteopetrosis in the bull syndrome of calcitonin excess. Am. J. clin. Nutr. 22: 115–118 (1969).
313 Kumar, M.A.; Foster, G.V.; MacIntyre, I.: Further evidence for calcitonin. A rapidly acting hormone which lowers plasma calcium. Lancet iii: 480 (1963).
314 Lambert, P.W.; Heath, H.H., III; Sizemore, G.W.: Pre- and postoperative studies of plasma calcitonin in primary hyperparathyroidism. J. clin. Invest. 63: 602–608 (1979).
315 Lambert, P.W.; Heath, H.; Sizemore, G.W.: Basal and stimulated plasma immunoreactive calcitonin (iCT) values are not high in primary hyperparathyroidism (1° HPT). Clin. Res. 24: 581A (1976).

References

316 Laskowski, J.: Carcinoma hyalinicum thyroideae. Nowotwory 7: 23–28 (1957).
317 Lawrence, A.M.: Pancreatic alpha-cell function in miscellaneous clinical disorders; in Glucagon, pp. 259–274 (Pergamon Press, Oxford 1972).
318 Leape, L.L.; Miller, H.H.; Graze, K.; Feldman, Z.T.; Gagel, R.F.; Wolfe, H.J.; DeLellis, R.A.; Tashjian, A.H., Jr.; Reichlin, S.: Total thyroidectomy for occult familial medullary carcinoma of the thyroid in children. J. pediat. Surg. 11: 831–837 (1976).
319 Leav, I.; Schiller, A.L.; Rijnberk, A.; Legg, M.A.; Kinderen, P.J. der: Adenomas and carcinomas of the canine and feline thyroid. Am. J. Path. 83: 61–93 (1976).
320 LeDourain, N.; LeLièvre, C.: Démonstration de l'origine neurale des cellules à calcitonine du corps ultimobranchial chez l'embryon de poulet. C.r. hebd. Séanc. Acad. Sci., Paris 270: 2857–2860 (1970).
321 Lee, J.C.; Catanzaro, A.; Parthemore, J.G.; Roach, B.; Deftos, L.J.: Hypercalcemia in disseminated coccidioidomycosis. New Engl. J. Med. 297: 431–433 (1977).
322 Lee, J.C.; Parthemore, J.G.; Deftos, L.J.: Immunochemical heterogeneity of calcitonin in renal failure. J. clin. Endocr. Metab. 45: 528–533 (1977).
323 Leicht, E.; Biro, G.; Weinges, K.F.: Inhibition of releasing hormone induced secretion of TSH and LH by calcitonin. Hormone metabol. Res. 6: 410–414 (1974).
324 Levy, M.; Habib, R.; Lyon, G.; Schweisguth, O.; Lemerle, J.; Roger, P.: Neuromatose et épithélioma à stroma amyloïde de la thyroïde chez l'enfant. Archs fr. Pédiat. 27: 561–583 (1970).
325 Liddle, G.W.; Island, D.P.; Ney, R.L.; Nicholson, W.E.; Shimizu, N.: Non-pituitary neoplasms and Cushing's syndrome. Ectopic 'adrenocorticotropin' produced by non-pituitary neoplasms as a cause of Cushing's syndrome. Archs intern. Med. 111: 471–475 (1963).
326 Liddle, G.W.; Givens, J.R.; Nicholson, W.E.; Island, D.P.: The ectopic ACTH syndrome. Cancer Res. 25: 1057–1061 (1965).
327 Linehan, W.M.; Farrell, R.E.; Cooper, C.W.; Wells, S.A., Jr.: Analysis of pentagastrin and calcium as thyrocalcitonin secretagogues in the early diagnosis of medullary carcinoma of the thyroid gland. Surg. Forum 28: 110–112 (1977).
328 Lips, K.J.; Sluys Veer, J. Van der; Struyvenberg, A.; Alleman, A.; Leo, J.R.; Wittebol, P.; Minder, W.H.; Kooiker, C.J.; Geerdink, R.A.; Waes, P.F. Van; Hackeng, W.H.: Bilateral occurrence of pheochromocytoma in patients with the multiple endocrine neoplasia syndrome, type 2A (Sipple's syndrome). Am. J. Med. 70: 1051–1060 (1981).
329 LiVolsi, V.A.; Feind, C.R.; LoGerfo, P.; Tashjian, A.H., Jr.: Demonstration by immunoperoxidase staining of hyperplasia of parafollicular cells in the thyroid gland in hyperparathyroidism. J. clin. Endocr. Metab. 37: 550–559 (1973).
330 Ljungberg, O.; Cederquist, E.; Studnitz, W. von: Medullary thyroid carcinoma and phaeochromocytoma. A familial chromaffinomatosis. Br. med. J. i: 279–281 (1967).
331 Ljungberg, O.: Argentaffin, parafollicular cells in the non-tumorous parenchyma of a thyroid gland with medullary thyroid carcinoma. Acta pathol. microbiol. scand. 78: 364–366 (1970).
332 Ljungberg, O.: Argentaffin cells in human thyroid and parathyroid and their relationship to C-cells and medullary carcinoma. Acta pathol. microbiol. scand. 80: 589–599 (1972).

333 Ljungberg, O.: On medullary carcinoma of the thyroid. A clinicopathologic entity. Acta pathol. microbiol. scand., suppl. 231, pp. 1–57 (1972).
334 Ljungberg, O.; Dymling, J.F.: Pathogenesis of C-cell neoplasia in thyroid gland. C-cell proliferation in a case of chronic hypercalcemia. Acta pathol. microbiol. scand. *80:* 577–588 (1972).
335 Loos, F.: Über doppelseitige Neurofibromatosis der Lider, Konjunctiva und Kornea, der Lippen und der Zunge. Klin. Mbl. Augenheilk. *89:* 184–189 (1932).
336 Lupulescu, A.: Effect of calcitonin on DNA synthesis in experimental wounds. Proc. Soc. exp. Biol. Med. *150:* 703–706 (1975).
337 MacGillivray, J.B.; Anderson, C.J.B.: Medullary carcinoma of the thyroid with parathyroid adenoma and hypercalcemia. J. clin. Path. *24:* 851–855 (1971).
338 Maier, R.; Kamber, B.; Riniker, B.; Rittel, W.: Analogues of human calcitonin. II. Influence of modifications in amino acid positions 1, 8 and 22 on hypocalcemic activity in rat. Hormone metabol. Res. *7:* 511–514 (1975).
339 Mandelstam, P.; Rush, B.F.; Mabry, C.C.; Bartlett, R.C.: Prophylactic thyroidectomy in a 4-½-year-old boy with a family history of oral and ocular mucous membrane neuromas, medullary carcinoma of the thyroid, pheochromocytoma, hyperparathyroidism, and diarrhea. J. Lab. clin. Med. *76:* 867 (1970).
340 Manning, P.C., Jr.; Molnar, G.D.; Black, B.M.; Priestley, J.T.; Woolner, L.B.: Pheochromocytoma, hyperparathyroidism, and thyroid carcinoma occurring coincidentally. New Engl. J. Med. *268:* 68–72 (1963).
341 Markey, W.S.; Ryan, W.G.; Economou, S.G.; Sizemore, G.W.; Arnaud, C.D.: Familial medullary carcinoma and parathyroid adenoma without pheochromocytoma. Ann. intern. Med. *78:* 898–901 (1973).
342 Marks, A.D.; Channick, B.J.: Extra-adrenal pheochromocytoma and medullary thyroid carcinoma with pheochromocytoma. Archs intern. Med. *134:* 1106–1109 (1974).
343 Marx, S.J.; Aurbach, G.D.: Renal receptors for calcitonin. Coordinate occurrence with calcitonin-activated adenylate cyclase. Endocrinology *97:* 448–453 (1975).
344 Matuchansky, D.; Bernier, J.J.: Effects of prostaglandin E on net and undirectional movements of water and electrolytes across the jejunal in man. Gut *12:* 854–855 (1971).
345 McKenna, T.J.; McLean, G.; Lorber, D.L.; Bone, H.G.; Parthemore, J.G.; Deftos, L.J.: Comparison of calcitonin stimulation tests used in screening for medullary carcinoma of the thyroid. Proc. 60th Meet. Endocrine Soc. A590 (1978).
346 McMillan, P.J.; Hooker, W.M.; Deftos, L.J.: Distribution of calcitonin-containing cells in human thyroid. Am. J. Anat. *140:* 73–80 (1974).
347 Meloni, C.R.; Tucci, J.; Canary, J.J.; Kyle, L.H.: Cushing's syndrome due to bilateral adrenocortical hyperplasia caused by a benign adrenal medullary tumor. J. clin. Endocr. Metab. *26:* 1192–1200 (1966).
348 Melvin, K.E.W.; Tashjian, A.H., Jr.: The syndrome of excessive thyrocalcitonin produced by medullary carcinoma of the thyroid. Proc. natn. Acad. Sci. USA *59:* 1216–1222 (1968).
349 Melvin, K.E.W.; Tashjian, A.H., Jr.; Cassidy, C.E.; Givens, J.R.: Cushing's syndrome caused by ACTH- and calcitonin-secreting medullary carcinoma of the thyroid. Metabolism *19:* 831–838 (1970).

350 Melvin, K.E.W.; Voelkel, E.F.; Tashjian, A.H., Jr.: Medullary carcinoma of the thyroid. Stimulation by calcium and glucagon of calcitonin secretion; in Calcitonin 1969, pp. 487–496 (Springer, New York 1970).
351 Melvin, K.E.W.; Miller, H.H.; Tashjian, A.H., Jr.: Early diagnosis of medullary carcinoma of the thyroid glands by means of calcitonin assay. New Engl. J. Med. *285:* 1115–1120 (1971).
352 Melvin, K.E.W.; Tashjian, A.H., Jr.; Miller, H.H.: Studies in familial (medullary) thyroid carcinoma. Recent Prog. Horm. Res. *28:* 399–470 (1972).
353 Melvin, K.E.W.; Tashjian, A.H., Jr.; Bordier, P.: The metabolic significance of calcitonin-secreting thyroid carcinoma; in clinical aspects of metabolic bone disease, pp. 193–201 (Excerpta Medica, Amsterdam 1973).
354 Melvin, K.E.W.: The paraneoplastic syndromes associated with carcinoma of the thyroid gland. Ann. N. Y. Acad. Sci. *230:* 378–390 (1974).
355 Melvin, K.E.W.: Medullary carcinoma of the thyroid. Pharmacol. Ther. *1:* 183–205 (1976).
356 Ménage, J.J.; Besnard, J.C.; Guilmot, J.L.; Vandooren, M.; Neel, J.L.: Preuves de l'absence de sécrétion de prostaglandines par un carcinome médullaire de la thyroïde avec diarrhée motrice. Nouv. Presse méd. *4:* 2862–2864 (1975).
357 Mendelsohn, G.; Baylin, S.B.; Eggleston, J.C.: Relationship of metastatic medullary thyroid carcinoma to calcitonin content of pheochromocytomas. An immunohistochemical study. Cancer *45:* 498–502 (1980).
358 Metz, S.A.; Deftos, L.J.; Baylink, D.J.; Robertson, R.P.: Neuroendocrine modulation of calcitonin and parathyroid hormone in man. J. clin. Endocr. Metab. *47:* 151–159 (1978).
359 Meyer, J.S.; Abdel-Bari, W.: Granules and thyrocalcitonin-like activity in medullary carcinoma of the thyroid. New Engl. J. Med. *278:* 523–529 (1968).
360 Mielke, J.E.; Becker, K.L.; Gross, J.B.: Diverticulitis of the colon in a young man with Marfan's syndrome associated with carcinoma of the thyroid gland and neurofibromas of the tongue and lips. Gastroenterology *48:* 379–382 (1965).
361 Milhaud, G.; Moukhtar, M.S.: Thyrocalcitonin. Effects on calcium kinetics in the rat. Proc. Soc. exp. Biol. Med. *123:* 207–209 (1966).
362 Milhaud, G.; Tubiana, M.; Parmentier, C.; Coutris, G.: Epithélioma de la thyroïde sécrétant de la thyrocalcitonine. C.r. hebd. Séanc. Sci., Paris *266:* 608–610 (1968).
363 Milhaud, G.; Calmettes, C.; Raymond, J.P.; Bignon, J.; Moukhtar, M.S.: Carcinoïde sécrétant de la thyrocalcitonine. C.r. hebd. Séanc. Acad. Sci., Paris *270:* 2195–2198 (1970).
364 Milhaud, G.; Calmettes, C.; Jullienne, A.; Tharaud, D.; Bloch-Michel, H.; Cavaillon, J.P.; Colin, R.; Moukhtar, M.S.: A new chapter in human pathology. Calcitonin disorders and therapeutic use; in Calcium, parathyroid hormone, and the calcitonins, pp. 56–70 (Excerpta Medica, Amsterdam 1972).
365 Milhaud, G.; Talbot, J.N.; Coutris, G.: Calcitonin treatment of postmenopausal osteoporosis. Evaluation of efficacy by principal components analysis. Biomedicine *22:* 223–232 (1975).
366 Milhaud, G.: Calcitonin 1975. Biomedicine *24:* 159–161 (1976).
367 Milhaud, G.; Moukhtar, M.S.: Thyrocalcitonin. Effects on calcium kinetics in the rat. Proc. Soc. exp. Biol. Med. *123:* 207–209 (1966).

368 Miller, R.L.; Burzynski, N.J.; Giammara, B.L.: The ultrastructure of oral neuromas in multiple mucosal neuromas, pheochromocytoma, medullary thyroid carcinoma syndrome. J. oral Pathol. *6:* 253–263 (1977).
369 Miravet, L.; Queille, M.L.; Carre, M.; Bordier, P.; Redel, J.: Action of vitamin D metabolites on the bone of vitamin D-deficient rats. Annls Biol. anim. Biochim. Biophys. *18:* 187–194 (1978).
370 Misiewicz, J.J.; Waller, S.L.; Kiley, N.: Effects of prostaglandin E_1 on intestinal transit in man. Lancet *i:* 648 (1969).
371 Moertel, C.G.; Beahrs, O.H.; Woolner, L.B.; Tyce, G.M.: 'Malignant carcinoid syndrome' associated with noncarcinoid tumors. New Engl. J. Med. *273:* 244–248 (1965).
372 Montalbano, F.P.; Baronofsky, I.D.; Bell, H.: Hyperplasia of the adrenal medulla. J. Am. med. Ass. *182:* 264–267 (1962).
373 Moorhead, E.L., II; Caldwell, J.R.; Kelly, A.R.; Morales, A.R.: The diagnosis of pheochromocytoma, analysis of 26 cases. J. Am. med. Ass. *196:* 1107–1113 (1966).
374 Morita, R.; Funkunaga, I.; Yamamoto, I.; Mori, T.; Torizuka, K.: Radioimmunoassay for human calcitonin employing synthetic calcitonin M. Its clinical application. Endocr. jap. *22:* 419–426 (1975).
375 Munzenberg, K.J.: Treatment of Sudeck's syndrome with calcitonin. Dt. med. Wschr. *103:* 26–29 (1978).
376 Nankin, H.; Hydovitz, J.; Sapira, J.: Normal chromosomes in mucosal neuroma variant of medullary thyroid carcinoma syndrome. J. med. Genet. *7:* 374–378 (1970).
377 Neher, R.; Riniker, B.; Maier, R.; Byfield, P.G.H.; Gudmundsson, T.V.; MacIntyre, I.: Human calcitonin. Nature, Lond. *220:* 984–986 (1968).
378 Neher, R.; Riniker, B.; Rittel, W.; Zuber, H.: Menschliches Calcitonin. III. Strukturen von Calcitonin M und D. Helv. chim. Acta *51:* 1900–1905 (1968).
379 Netzloff, M.L.; Garnica, A.D.; Rodgers, B.M.; Firas, J.L.: Medullary carcinoma of the thyroid in the multiple mucosal neuromas syndrome. Ann. clin. Lab. Sci. *9:* 368–373 (1979).
380 Niall, H.D.; Keutmann, H.T.; Copp, D.H.; Potts, J.T., Jr.: Amino acid sequence of salmon ultimobranchial calcitonin. Proc. natn. Acad. Sci. USA *64:* 771–778 (1969).
381 Noda, T.; Narita, K.: Amino acid sequence of eel calcitonin. J. Biochem., Tokyo *79:* 353–359 (1976).
382 Nonidez, J.F.: The origin of the 'parafollicular' cell, a second epithelial component of the thyroid gland in the dog. Am. J. Anat. *49:* 479–505 (1932).
383 Norberg, H.P.; DeRoos, J.; Kaplan, E.L.: Increased parathyroid hormone secretion and hypocalcemia in experimental pancreatitis. Necessity for an intact thyroid gland. Surgery, St. Louis *77:* 773–779 (1975).
384 Norman, T.; Gautvik, K.M.; Johannessen, J.V.; Brennhovd, I.O.: Medullary carcinoma of the thyroid in Norway. Acta endocr., Copenh. *83:* 71–85 (1976).
385 Normann, T.; Johannessen, J.V.; Gautvik, K.M.; Olsen, B.R.; Brennhovd, I.O.: Medullary carcinoma of the thyroid, diagnostic problems. Cancer *38:* 366–377 (1976).
386 Normann, T.: Medullary thyroid carcinoma in Norway. Epidemiological and genetic data. Acta pathol. microbiol. scand. *85:* 775–786 (1977).

387 Normann, T.; Gautvik, K.M.: Medullary carcinoma of the thyroid gland in Norway. Serum calcitonin in 300 relatives of 43 patients. Annls Chir. Gynaec. Fenn. *66:* 187–194 (1977).
388 Norton, J.A.; Froome, L.C.; Farrell, R.E.; Wells, S.A., Jr.: Multiple endocrine neoplasia type IIb. The most aggressive form of medullary thyroid carcinoma. Surg. Clins N. Am. *59:* 109–118 (1979).
389 Nourok, D.S.: Familial pheochromocytoma and thyroid carcinoma. Ann. intern. Med. *60:* 1028–1040 (1964).
390 Nozaki, K.; Noda, S.; Obi, S.; Nishizawa, Y.; Morii, H.; Wada, M.: Secretion of gastrin and calcitonin after ingestion of meat extract in patients with peptic ulcer. Endocr. jap. *23:* 83–86 (1976).
391 Nunez, E.A.; Gould, R.P.; Holt, S.J.: Autophagy of unusual parafollicular ('C') cell granules in bat thyroid glands; in Taylor, Calcitonin 1969. Proc. 2nd int. Symp., London 1969, pp. 252–261 (Heinemann, London 1970).
392 Olson, E.B., Jr.; DeLuca, H.F.; Potts, J.T., Jr.: The effect of calcitonin and parathyroid hormone on calcium transport of isolated intestine; in Talmage, Munson, Calcium, parathyroid hormone, and the calcitonins. Proc. 4th Parathyroid Conf., Chapel Hill, N.C. 1971, pp. 240–246 (Excerpta Medica, Amsterdam 1972).
393 Olson, B.E., Jr.; DeLuca, H.F.; Potts, J.T., Jr.: Calcitonin inhibition of vitamin D-induced intestinal calcium absorption. Endocrinology *90:* 151–157 (1972).
394 Otani, M.; Yamauchi, H.; Meguro, T.; Kitazawa, S.; Watanabe, S.; Orimo, H.: Isolation and characterization of calcitonin from pericardium and esophagus of eel. J. Biochem., Tokyo *79:* 345–352 (1976).
395 Oubert, I.; Pedinielli, L.; Carlopino, C.; Detrolle, P.; Sudaka, P.; Armand, P.: Phéochromocytome bilatéral avec tension artérielle basse et diabète sucré. Bull. Soc. méd. Hôp., Paris *115:* 891–903 (1964).
396 Pages, A.; Marty, C.H.; Baldet, P.; Peraldi, R.: Le syndrome neurofibromatose – carcinome médullaire thyroïdien – phéochromocytome. Archs Anat. path. *18:* 137–142 (1970).
397 Pak, C.Y.C.; Steward, A.; Kaplan, R.; Bone, H.; Notz, C.; Browne, R.: Photon absorptiometric analysis of bone density in primary hyperparathyroidism. Lancet *ii:* 7–8 (1975).
398 Paloyan, E.; Paloyan, D.; Harper, P.V.: The role of glucagon hypersecretion in the relationship of pancreatitis and hyperparathyroidism. Surgery, St. Louis *62:* 167–173 (1967).
399 Parthemore, J.G.; Bronzert, D.; Roberts, G.; Deftos, L.J.: A short calcium infusion in the diagnosis of medullary thyroid carcinoma. J. clin. Endocr. Metab. *39:* 108–111 (1974).
400 Parthemore, J.G.; Deftos, L.J.: The regulation of calcitonin in normal human plasma as assessed by immunoprecipitation and immunoextraction. J. clin. Invest. *56:* 835 (1975).
401 Parthemore, J.G.; Deftos, L.J.: Calcitonin secretion in normal human subjects. J. clin. Endocr. Metab. *47:* 184–188 (1978).
402 Parthemore, J.G.; Deftos, L.J.: Secretion of calcitonin in primary hyperparathyroidism. J. clin. Endocr. Metab. *49:* 223–226 (1979).
403 Paterson, C.R.: Vitamin D resistance in hypoparathyroidism with medullary carcinoma of the thyroid. Br. med. J. *i:* 952 (1977).

404 Patnaik, A.D.; Lieberman, P.H.; Erlandson, R.A.; Acevedo, W.M.; Liu, S.K.: Canine medullary carcinoma of the thyroid. Vet. Pathol. *15:* 590–599 (1978).
405 Pearse, A.G.E.: The cytochemistry of the thyroid cells and their relationship to calcitonin. Proc. R. Soc. *164:* 478–487 (1966).
406 Pearse, A.G.E.: 5-Hydroxytryptophan uptake by dog thyroid C-cells and its possible significance in polypeptide hormone production. Nature, Lond. *211:* 598–600 (1966).
407 Pearse, A.G.E.: Common cytochemical properties of cells producing polypeptide hormones with particular reference to calcitonin and the thyroid C-cells. Vet. Rec. *79:* 587–590 (1966).
408 Pearse, A.G.E.: Common cytochemical and ultrastructural characteristics of cells producing polypeptide hormones (the APUD series) and their relevance to thyroid and ultimobranchial C-cells and calcitonin. Proc. R. Soc. *170:* 71–80 (1968).
409 Pearse, A.G.E.; Polak, J.M.: Cytochemical evidence for the neural crest origin of mammalian ultimobranchial cells. Histochemie *27:* 96–102 (1971).
410 Pearse, A.G.E.; Ewen, S.E.B.; Polak, J.M.: The genesis of APUD amyloid in endocrine polypeptide tumors. Histochemical distinction from immunamyloid. Virchows Arch. Abt. B Zellpath. *10:* 93–107 (1972).
411 Pearse, A.G.E.; Polak, J.M.; Noorden, S. van: The neural crest origin of the C-cells and their comparative cytochemistry and ultrastructure in the ultimobranchial gland; in Talmage, Munson, Calcium, parathyroid hormone, and the calcitonins. Proc. 4th Parathyroid Conf., Chapel Hill, N.C. 1971, pp. 29–40 (Excerpta Medica, Amsterdam 1972).
412 Pearson, K.D.; Wells, S.A.; Keiser, H.R.: Familial medullary carcinoma of the thyroid, adrenal pheochromocytoma, and parathyroid hyperplasia. Radiology *107:* 249–256 (1973).
413 Pecile, A.; Ferri, S.; Braga, P.C.; Olgiati, V.R.: Effects of intracerebroventricular calcitonin in the conscious rabbit. Experientia *31:* 332–333 (1975).
414 Pento, J.T.; Glick, S.M.; Kagen, A.; Gorfein, P.C.: The relative influence of calcium, strontium, and magnesium on calcitonin secretion in the pig. Endocrinology *94:* 1176–1180 (1974).
415 Perlow, M.J.; Freed, W.J.; Carman, J.S.; Wyatt, R.J.: Calcitonin reduces feeding in man, monkey, and rat. Pharmacol. Biochem. Behav. *12:* 609–612 (1980).
416 Pitkin, M.; Reynolds, W.A.; Williams, G.A.; Hargis, G.K.: Calcium-regulating hormones during the menstrual cycle. J. clin. Endocr. Metab. *47:* 626–632 (1978).
417 Polak, J.M.; Pearse, A.G.E.; LeLievre, C.; Fontaine, J.; LeDourain, N.M.: Immunocytochemical confirmation of the neural crest origin of avian calcitonin-producing cells. Histochemistry *40:* 209–214 (1974).
418 Popovtzer, M.M.; Blum, M.S.; Flis, R.S.: Evidence for interference of 25(OH) vitamin D_3 with phosphaturic action of calcitonin. Am. J. Physiol. E *232:* 515–521 (1977).
419 Potts, J.T., Jr.; Niall, H.D.; Deftos, L.J.: Calcitonin; in Martini, James, Current topics in experimental endocrinology, pp. 151–173 (Academic Press, New York 1971).
420 Putter, I.; Kacka, E.A.; Harman, R.E.; Rickes, E.; Kempf, A.J.; Chaiet, L.; Rothrock, J.W.; Wase, A.; Wolf, F.J.: Thyrocalcitonin. Isolation and chemical properties; in

Taylor, Calcitonin. Proc. Symp. Thyrocalcitonin and C-cells, pp. 74–76 (Heinemann, London 1968).

421 Raisz, L.G.; Au, W.Y.W.; Friedman, J.; Niemann, I.: Inhibition of bone resorption in tissue culture by thyrocalcitonin; in Taylor, Calcitonin. Proc. Symp. Thyrocalcitonin and C-cells, pp. 215–222 (Heinemann, London 1968).

422 Raker, J.W.; Henneman, P.H.; Graf, W.S.: Co-existing primary hyperparathyroidism and Cushing's syndrome. J. clin. Endocr. 22: 273–289 (1962).

423 Rambaud, J.C.; Nisard, A.; Modigliani, R.; Calmette, C.; Moukhtar, M.S.; Hill, P.A.; Besterman, H.: Hypercalcitoninaemia in vipomas. Lancet i: 220 (1978).

424 Rappaport, H.M.: Neurofibromatosis of the oral cavity. Report of a case. Oral Surg. 6: 599–604 (1953).

425 Rasmussen, B.: Magnesium and phosphate in the serum of patients with medullary carcinoma of the thyroid. Clinica chim. Acta 89: 279–283 (1978).

426 Rasmussen, H.; Wong, M.; Bikle, D.; Goodman, D.B.P.: Hormonal control of the renal conversion of 25-hydroxycholecalciferol to 1,25-dihydroxycholecalciferol. J. clin. Invest. 51: 2502–2504 (1972).

427 Raulais, D.; Hagaman, J.; Ontjes, D.A.; Lundblad, R.L.; Kingdon, H.S.: The complete amino-acid sequence of rat thyrocalcitonin. Eur. J. Biochem. 64: 607–611 (1976).

428 Ravid, M.; Gafni, J.; Sohar, E.; et al.: Incidence and origin of non-systemic microdeposits of amyloid. J. clin. Path. 20: 15–20 (1967).

429 Raynor, A.C.; Sowden, D.: Medullary thyroid carcinoma. J. surg. Oncol. 7: 435–445 (1975).

430 Riggs, B.L.; Jowsey, J.; Kelly, P.J.; Arnaud, C.D.: Role of hormonal factors in the pathogenesis of postmenopausal osteoporosis. Israel J. med. Sci. 12: 615–619 (1976).

431 Riniker, B.; Neher, R.; Maier, R.; Kahnt, F.W.; Byfield, P.G.H.; Gudmundsson, T.V.; Galante, L.; MacIntyre, I.: Menschliches Calcitonin. I. Isolierung und Charakterisierung. Helv. chim. Acta 51: 1738–1742 (1968).

432 Rittel, W.; Brugger, M.; Kamber, B.; Riniker, B.; Sieber, P.: Thyrocalcitonin. III. Die Synthese des Alpha-Thyrocalcitonins. Helv. chim Acta 51: 924–928 (1968).

433 Rittel, W.; Maier, R.; Brugger, M.; Kamber, B.; Riniker, B.; Sieber, P.: Structure-activity relationships of human calcitonin. III. Biological activity of synthetic analogues with shortened or terminally modified peptide chains. Experientia 32: 246–248 (1976).

434 Robertson, G.M.; Sizemore, G.W.; Gordon, H.: Thickened corneal nerves as a manifestation of multiple endocrine neoplasia. Trans. Am. Acad. Ophthal. Oto-lar. 70: 772–787 (1975).

435 Robertson, G.M., Jr.; Moore, B.W.; Switz, D.M.; Sizemore, G.W.; Estep, H.L.: Inadequate parathyroid response in acute pancreatitis. New Engl. J. Med. 294: 512–516 (1976).

436 Robinson, W.W.: Need for further study as to the role of the parathyroids in Cushing's syndrome. J. clin. Endocr. 14: 811 (1954).

437 Roediger, W.E.: Thyroidectomy for non-familial medullary carcinoma. Br. J. Surg. 63: 343–345 (1976).

438 Roos, B.A.; Okano, K.; Deftos, L.J.: Evidence for a pro-calcitonin. Biochem. biophys. Res. Commun. 60: 1134–1140 (1974).

439 Roos, B.A.; Deftos, L.J.: Radioimmunoassay of calcitonin in plasma, normal thyroid, and medullary thyroid carcinoma of the rat. J. Lab. clin. Med. 88: 173–182 (1976).
440 Roos, B.A.; Parthemore, J.G.; Lee, J.C.; Deftos, L.J.: Calcitonin heterogeneity. In vivo and in vitro studies. Calcif. Tissue Res. 22S: 298–302 (1977).
441 Roos, B.A.; Cooper, C.W.; Frelinger, A.L.; Deftos, L.J.: Acute and chronic fluctuations of immunoreactive and biologically active plasma calcitonin in the rat. Endocrinology 103: 2180–2186 (1978).
442 Roos, B.A.; Lindall, A.W.; Ellis, J.; Eide, R.; Lambert, P.W.; Birnbaum, R.S.: Increased plasma and tumor somatostatin-like immunoreactivity in medullary thyroid carcinoma and small cell lung cancer. J. clin. Endocr. Metab. 52: 187–194 (1981).
443 Rosenberg, E.M.; Hahn, T.J.; Orth, D.N.; Deftos, L.J.; Tanaka, K.: ACTH-secreting medullary carcinoma of the thyroid presenting as severe idiopathic osteoporosis and senile purpura. Report of a case and review of the literature. J. clin. Endocr. Metab. 47: 255–262 (1978).
444 Rude, R.K.; Singer, F.R.: Comparison of serum levels after a one-minute calcium injection and after pentagastrin injection in diagnosis of medullary thyroid carcinoma. J. clin. Endocr. Metab. 44: 980–983 (1977).
445 Ruppert, R.D.; Buerger, L.F.; Chang, W.W.L.: Pheochromocytoma, neurofibromatosis, and thyroid carcinoma. Metabolism 15: 537–541 (1966).
446 Saharia, P.C.: Carcinoma arising in thyroglossal duct remnant. Case reports and review of the literature. Br. J. Surg. 62: 689–691 (1975).
447 Samaan, N.A.; Hill, C.S., Jr.; Beceiro, J.R.; Schultz, P.N.: Immunoreactive calcitonin in medullary carcinoma of the thyroid and in maternal and cord serum. J. Lab. clin. Med. 81: 671–681 (1973).
448 Samaan, N.A.; Wigoda, C.; Castillo, S.G.: Human serum calcitonin and parathyroid hormone levels in the maternal umbilical cord blood and postpartum; in Taylor, Endocrinology 1973. Proc. 4th int. Symp., pp. 364–372 (Heinemann, London 1974).
449 Samaan, N.A.; Anderson, G.D.; Adam-Mayne, M.E.: Immunoreactive calcitonin in the mother, neonate, child, and adult. Am. J. Obstet. Gynec. 121: 622–625 (1975).
450 Samaan, N.A.; Ibanez, M.; Hill, C.S., Jr.: Medullary carcinoma of the thyroid and astrocytoma. Ann. intern. Med. 86: 585–586 (1977).
451 Sarosi, G.; Doe, R.P.: Familial occurrence of parathyroid adenomas, pheochromocytoma, and medullary carcinoma of the thyroid with amyloid stroma (Sipple's syndrome). Ann. intern. Med. 68: 1305–1309 (1968).
452 Sauer, R.; Niall, H.D.; Potts, J.T., Jr.: Accelerated procedures for automated peptide degradation. Fed. Proc. 29: 782/2723 (1970).
453 Schimke, R.N.; Hartmann, W.H.: Familial amyloid-producing medullary thyroid carcinoma and pheochromocytoma. A distinct genetic entity. Ann. intern. Med. 63: 1027–1039 (1965).
454 Schimke, R.N.; Hartmann, W.H.; Prout, T.E.; Rimoin, D.L.: Syndrome of bilateral pheochromocytoma, medullary thyroid carcinoma, and multiple neuromas. New Engl. J. Med. 279: 1–7 (1968).
455 Schimke, R.N.: Multiple mucosal neuromata syndrome; in Lynch, Skin, Heredity

and malignant neoplasma, pp. 208–219 (Medical Examination Publishing, New York 1972).

456 Schlemmer, R.F.; Farnsworth, N.R.; Cordell, G.A.; Bederka, J.P.: Nutmeg pharmacognosy. New Engl. J. Med. *289:* 922 (1973).
457 Schocket, E.; Teloh, H.A.: Aganglioma megacolon, pheochromocytoma, megaloureter and neurofibroma. Co-occurrence of several neural abnormalities. Am. J. Dis. Child. *94:* 185–191 (1957).
458 Seyberth, H.W.; Segre, G.V.; Morgan, J.L.; Sweetman, B.J.; Potts, J.T., Jr.; Oates, J.A.: Prostaglandins as mediators of hypercalcemia associated with cancer. New Engl. J. Med. *293:* 1278–1283 (1975).
459 Shamonki, I.M.; Frumar, A.M.; Tataryn, I.V.; Meldrum, D.R.; Davidson, B.H.; Parthemore, J.G.; Judd, H.L.; Deftos, L.J.: Age-related changes of calcitonin secretion in females. J. clin. Endocr. Metab. (in press).
460 Sherwin, R.P.: Present status of the pathology of the adrenal gland in hypertension. Am. J. Surg. *107:* 136–143 (1964).
461 Shibaski, T.; Deftos, L.; Guillemin, R.: Immunoreactive-β-endorphin, -adrenocorticotropin, and -calcitonin in extracts of anaplastic or differentiated (rat) medullary thyroid carcinoma. Biochem. biophys. Res. Commun. *90:* 1266–1273 (1979).
462 Shimaoka, K.: Adjunctive management of thyroid cancer. Chemotherapy. J. surg. Oncol. *15:* 283–286 (1980).
463 Sieber, P.; Brugger, M.; Kamber, B.; Riniker, B.; Rittel, W.: Menschliches Calcitonin. IV. Die Synthese von Calcitonin M. Helv. chim. Acta *51:* 2057–2061 (1968).
464 Sieber, P.; Brugger, M.; Kamber, B.; Riniker, B.; Rittel, W.; Maier, R.; Staehelin, M.: Synthesis and biological activity of peptide sequences related to porcine a-thyrocalcitonin; in Taylor, Calcitonin 1969. Proc. 2nd int. Symp., London 1969, pp. 28–33 (Heinemann, London 1970).
465 Silva, O.L.; Becker, K.L.; Primack, A.; Doppman, J.; Snider, R.H.: Ectopic secretion of calcitonin by oat-cell carcinoma. New Engl. J. Med. *290:* 1122–1124 (1974).
466 Silva, O.L.; Becker, K.L.; Selawry, H.P.; Snider, R.H.; Moore, C.F.; Bivins, L.E.; Shalgoub, R.J.: Human calcitonin and serum-phosphate. Lancet *i:* 1055 (1974).
467 Silva, O.L.; Becker, K.L.; Doppman, J.L.; Snider, R.H.; Moore, C.F.: Calcitonin levels in thyroid-vein blood of man. Am. J. med. Sci. *269:* 37–41 (1975).
468 Silva, O.L.; Snider, R.H.; Becker, K.L.; Moore, C.F.: Immunochemical forms of calcitonin in man, clinical implications. Fed. Proc. *34:* 761 (1975).
469 Silva, O.L.; Becker, K.L.: High plasma calcitonin levels in breast cancer. Br. med. J. *i:* 460 (1976).
470 Silva, O.L.; Becker, K.L.; Primack, A.; Doppman, J.L.; Snider, R.H.: Increased serum calcitonin levels in bronchogenic cancer. Chest *69:* 495–499 (1976).
471 Silva, O.L.; Wisneski, L.A.; Cyrus, J.; Snider, R.H.; Moore, C.F.; Becker, K.L.: Calcitonin in thyroidectomized patients. Am. J. med. Sci. *275:* 159–164 (1978).
472 Simpson, H.E.: Oral neurofibromatosis with differentiation of sensory organs. Oral Surg. *19:* 228–233 (1965).
473 Singer, F.R.; Aldred, J.P.; Neer, R.M.; Krane, S.M.; Potts, J.T., Jr.; Bloch, K.J.: An evaluation of antibodies and clinical resistance to salmon calcitonin. J. clin. Invest. *51:* 2331–2338 (1972).
474 Singer, F.R.; Habener, J.F.; Greene, E.; Godin, P.; Potts, J.T., Jr.: Inactivation of calcitonin by specific organs. Nature, Lond. *237:* 269–270 (1972).

475 Singer, F.R.; Habener, J.F.: Multiple immunoreactive forms of calcitonin in human plasma. Biochem. biophys. Res. Commun. *61:* 660–666 (1974).
476 Sipple, J.H.: The association of pheochromocytoma with carcinoma of the thyroid gland. Am. J. Med. *31:* 163–166 (1961).
477 Sizemore, G.W.; Go, V.L.W.; Kaplan, E.L.; Sanzenbacher, L.J.; Holtermulller, K.H.; Arnaud, C.D.: Relations of calcitonin and gastrin in the Zollinger-Ellison syndrome and medullary carcinoma of the thyroid. New Engl. J. Med. *288:* 641–644 (1973).
478 Sizemore, G.W.; Heath, H.: Immunochemical heterogeneity of calcitonin in plasma of patients with medullary thyroid carcinoma. J. clin. Invest. *55:* 111–118 (1975).
479 Sizemore, G.W.; Carney, J.A.; Heath, H.: Epidemiology of medullary carcinoma of thyroid gland – 5-year experience (1971–1976). Surg. Clins N. Am. *57:* 633–646 (1977).
480 Sjoerdsma, A.; Engleman, K.; Waldman, T.A.; Cooperman, L.H.; Hammond, W.G.: Pheochromocytoma. Current concepts of diagnosis and treatment. Ann. intern. Med. *65:* 1302–1326 (1966).
481 Skrabanek, P.; Cannon, D.; Dempsey, J.; Kirrane, J.; Neligan, M.; Powell, D.: Substance P in medullary carcinoma of the thyroid. Experientia *35:* 1295–1260 (1979).
482 Sletten, K.; Westermark, P.; Natvig, J.B.: Characterization of amyloid fibril proteins from medullary carcinoma of the thyroid. J. exp. Med. *143:* 993–998 (1976).
483 Smiths, M.; Huizinga, J.: Familial occurrence of phaeochromocytoma. Acta genet. *11:* 137–153 (1961).
484 Snider, R.H.; Silva, O.L.; Moore, C.F.; Becker, K.L.: Immunochemical heterogeneity of calcitonin in man. Effect on radioimmunoassay. Clinica chim. Acta *76:* 1–14 (1977).
485 Sowa, M.; Appert, H.E.; Howard, J.M.: Hypocalcemic activity of pancreatic tissue homogenate in dog. Surgery Gynec. Obstet. *144:* 365–370 (1977).
486 Spiler, I.J.; Kapcala, L.P.; Graze, K.; Gagel, R.F.; Feldman, Z.T.; Biller, B.; Tashjian, A.H., Jr.; Reichlin, S.: Effects of *l*-dopa and bromocriptine on calcitonin secretion in medullary thyroid carcinoma. J. clin. Endocr. Metab. *51:* 806–809 (1980).
487 Staehelin, M.: The calcitonins. An example of unusual evolution. J. mol. Evol. *1:* 258–262 (1972).
488 Steiner, A.L.; Goodman, A.D.; Powers, S.R.: Study of a kindred with pheochromocytoma, medullary thyroid carcinoma, hyperparathyroidism, and Cushing's disease: MEN, type II. Medicine, Baltimore *47:* 371–409 (1968).
489 Stepanas, A.V.; Samaan, N.A.; Hill, C.S., Jr.; Hickey, R.C.: Medullary thyroid carcinoma – importance of serial serum calcitonin measurement. Cancer *43:* 825–827 (1979).
490 Stjernholm, M.R.; Freudenbourg, J.C.; Mooney, H.S.; Kinney, F.J.; Deftos, L.J.: Medullary carcinoma of the thyroid at 18 months of age. New Engl. J. Med. (in press).
491 Stoffel, E.: Lokales Amyloid der Schilddrüse. Virchows Arch. path. Anat. Physiol. *201:* 245–252 (1910).
492 Strauss, E.; Yalow, R.S.: Artifacts in the radioimmunoassay of peptide hormones in gastric and duodenal secretions. J. Lab. clin. Med. *87:* 292–298 (1976).
493 Strettle, R.J.; Bates, R.F.L.; Buckley, G.A.: Evidence for a direct anti-inflammatory action of calcitonin. Inhibition of histamine-induced mouse Pinnal oedema by porcine calcitonin. J. Pharm. Pharmac. *32:* 192–195 (1980).

494 Sundler, F.; Aluments, J.; Hakanson, R.; Bjorklund, L.; Ljungberg, O.: Somatostatin-immunoreactive cells in medullary carcinoma of the thyroid. Am. J. Path. 88: 381–384 (1977).
495 Swenson, O.; Fisher, J.H.: The relation of megacolon and megaloureter. New Engl. J. Med. 253: 1147–1150 (1955).
496 Swinton, N.W.; Clerkin, E.P.; Flint, L.D.: Hypercalcemia and familial pheochromocytoma. Correction after adrenalectomy. Ann. intern. Med. 76: 455–456 (1972).
497 Takaoka, Y.; Takamori, M.; Ichinose, M.; Shikaya, T.; Igawa, N.; Kikutani, M.; Yamamoto, Y.: Hypocalcemic action of a pancreatic factor and its clinical significance on the myasthenic patients. Acta med., Nagasaki 13: 28–35 (1969).
498 Tamburino, G.; Fiore, C.E.; Cottini, E.; Petralito, A.: Plasma calcitonin after calcium infusion in normal adults and in senile osteoporosis. IRCS med. Sci. 4: 41 (1976).
499 Tasca, C.; Stefaneanu, L.: Histopathologic observations on thyroid carcinoma with amyloid stroma. Morphol. Embryol. 21: 103–107 (1975).
500 Tashjian, A.H., Jr.; Voelkel, E.F.: Decreased thyrocalcitonin in thyroid glands from patients with hyperparathyroidism. J. clin. Endocr. Metab. 27: 1353–1357 (1967).
501 Tashjian, A.H., Jr.; Howland, B.G.; Kenneth, B.A.; Melvin, E.W.; Hill, C.S., Jr.: Immunoassay of human calcitonin. Clinical measurement, relation to serum calcium and studies in patients with medullary carcinoma. New Engl. J. Med. 283: 890–895 (1970).
502 Tashjian, A.H., Jr.; Wright, D.R.; Ivey, J.L.; Pont, A.: Calcitonin binding sites in bone. Relationships to biological response and 'escape'. Recent Prog. Horm. Res. 34: 285–287 (1978).
503 Taylor, T.G.; Lewis, P.E.; Balderstone, O.: Role of calcitonin in protecting the skeleton during pregnancy and lactation. J. Endocr. 66: 297–298 (1975).
504 Texter, E.C., Jr.: Hirschsprung's disease. Am. J. dig. Dis. 1: 35–46 (1956).
505 Thies, W.: Multiple echte fibrillare Neurone (Rankenneurone) der Haut und Schleimhaut. Arch. klin. exp. Derm. 218: 561–573 (1964).
506 Thiliveris, J.A.; Dube, W.J.; Banerjee, R.: Comparative ultrastructure of thyroid, tongue and eyelid lesions in the neuroma phenotype of medullary carcinoma of the thyroid. Association of amyloid with fibroblasts in thyroid tumor and in mucosal neuromas. Virchows Arch. Abt. A Path. Anat. 369: 249–258 (1976).
507 Tischler, A.S.; Dichter, M.A.; Biales, D.; Green, L.G.: Neuroendocrine neoplasma and their cells of origin. New Engl. J. Med. 296: 919–925 (1977).
508 Toverud, S.U.; Harper, C.; Munson, P.L.: Calcium metabolism during lactation. Enhanced effects of thyrocalcitonin. Endocrinology 99: 371–378 (1976).
509 Trump, D.L.; Livingston, N.J.; Baylin, S.B.: Water diarrhea syndrome in an adult with ganglioneuroma-pheochromocytoma – identification of vasoactive intestinal peptide, calcitonin, and catecholamines and assessment of their biologic activity. Cancer 40: 1526–1532 (1977).
510 Tubiana, M.; Milhaud, G.; Coutris, G.; Lacour, J.; Parmentier, C.; Bok, B.: Medullary carcinoma and thyrocalcitonin. Br. med. J. iv: 87–89 (1968).
511 Valenta, L.J.; Michelbechet, M.; Mattson, J.C.; Singer, F.R.: Microfollicular thyroid carcinoma with amyloid rich stroma, resembling medullary thyroid carcinoma of thyroid. Cancer 39: 1573–1587 (1977).
512 Valk, T.W.; Frager, M.S.; Gross, M.D.; Sisson, J.C.; Wieland, D.M.; Swanson, D.P.; Mangner, T.J.; Beierwaltes, W.H.: Spectrum of pheochromocytoma in multiple

endocrine neoplasia. A scintigraphic portrayal using ^{131}I-metaiodobenzylguanidine. Ann. intern. Med. *94:* 762–767 (1981).

513 Epps, E.F. Van; Hyndman, O.R.; Greene, J.A.: Clinical manifestations of paroxysmal hypertension associated with pheochromocytoma of adrenal. Report of a proved and of a doubtful case. Archs intern. Med. *65:* 1123–1129 (1940).

514 Velo, G.P.; DeBastiani, G.; Nogarin, L.; Abdullahi, S.E.: Antiinflammatory effect of calcitonin. Agents Actions *6:* 284 1976).

515 Verdy, M.; Beaulieu, R.; Demers, L.; Sturtridge, W.C.; Thomas, P.; Kumar, M.A.: Plasma calcitonin activity in a patient with thyroid medullary carcinoma and her children with osteopetrosis. J. clin. Endocr. Metab. *32:* 216–221 (1971).

516 Visser, J.W.; Axt, R.: Bilateral adrenal medullary hyperplasia. A clinicopathological entitiy. J. clin. Path. *28:* 298–304 (1975).

517 Voelkel, E.F.; Tashjian, A.H., Jr.; Davidoff, F.F.; Cohen, R.B.; Perlia, C.P.; Wurtman, R.J.: Concentrations of calcitonin and catecholamines in pheochromocytomas, a mucosal neuroma and medullary thyroid carcinoma. J. clin. Endocr. Metab. *37:* 297–307 (1973).

518 Vora, N.M.; Williams, G.A.; Hargis, G.K.; Bowser, E.N.; Kawahara, W.; Jackson, B.L.; Henderson, W.J.; Kukreja, S.C.: Comparative effect of calcium and of the adrenergic system on calcitonin secretion in man. J. clin. Endocr. Metab. *46:* 567–571 (1977).

519 Wahner, H.W.; Cuello, C.; Aljure, F.: Hormone-induced regression of medullary (solid) thyroid carcinoma. Am. J. Med. *45:* 789–794 (1968).

520 Walker, D.M.: Oral mucosal neuroma – medullary thyroid carcinoma syndrome. Br. J. Derm. *88:* 599–603 (1973).

521 Wallace, S.; Hill, C.S.; Paulus, D.D., Jr.; Ibanez, M.L.; Clark, R.L.: The radiologic aspects of medullary (solid) thyroid carcinoma. Radiol. Clin. N. Am. *8:* 463–474 (1970).

522 Wallach, S.R.; Royston, I.; Taetle, R.; Wohl, H.; Deftos, L.J.: Plasma calcitonin as a marker of disease activity in patients with small cell carcinoma of the lung. J. clin. Endocr. Metab. *53:* 602–606 (1981).

523 Watkins, W.B.; Moore, R.Y.; Burton, D.; Deftos, L.J.: Distribution of immunoreactive calcitonin in the rat pituitary gland. Endocrinology *106:* 1966–1970 (1980).

524 Watts, E.G.; Copp, D.H.; Deftos, L.J.: Changes in plasma calcitonin and calcium during the migration of salmon. Endocrinology *96:* 214–218 (1974).

525 Weichert, R.F., III: The neural ectodermal origin of the peptide-secreting endocrine glands. A unifying concept for the etiology of multiple endocrine adenomatosis and the inappropriate secretion of peptide hormones by nonendocrine tumors. Am. J. Med. *49:* 232–241 (1970).

526 Weiss, R.E.; Singer, F.R.; Gorn, A.H.; Hofer, D.P.; Nimni, M.E.: Calcitonin stimulates bone formation when administered prior to initiation of osteogenesis. J. clin. Invest. *68:* 815–818 (1981).

527 Weir, G.C.; Lesser, P.B.; Drop, L.J.; Fischer, J.E.; Warshaw, A.L.: The hypocalcemia of acute pancreatitis. Ann. intern. Med. *83:* 185–189 (1975).

528 Wells, S.A., Jr.; Ontjes, D.A.; Copper, C.W.; Hennessey, J.F.; Ellis, G.J.; MacPherson, H.T.; Sabiston, D.C., Jr.: The early diagnosis of medullary carcinoma of the thyroid gland in patients with multiple endocrine neoplasia, type II. Ann. Surg. *182:* 362–370 (1975).

529 Wells, S.A., Jr.; Ontjes, D.A.: Multiple endocrine neoplasia type II. Ann. Rev. Med. 27: 263–268 (1976).
530 Wells, S.A.; Haagensen, D.E.; Linehan, W.M.; Farrell, R.E.; Dilley, W.G.: Detection of elevated plasma levels of carcinoembryonic antigen in patients with suspected or established medullary thyroid carcinoma. Cancer 42: 1498–1503 (1978).
531 Werner, S.; Low, H.: Inhibitory effects of calcitonin on lipolysis and ^{47}calcium accumulation in rat adipose tissue in vivo. Hormone metabol. Res. 6: 30–36 (1974).
532 Westermark, P.: Amyloid of medullary carcinoma of the thyroid. Partial characterization. Uppsala J. med. Sci. 80: 88–92 (1975).
533 Whittle, T.S., Jr.; Goodwin, M.N., Jr.: Intestinal ganglioneuromatosis with the mucosal neuroma – medullary thyroid carcinoma – pheochromocytoma syndrome. Am. J. Gastroent., N.Y. 65: 249–257 (1976).
534 Williams, E.D.: A review of 17 cases of carcinoma of the thyroid and phaeochromocytoma. J. clin. Path. 18: 288–292 (1965).
535 Williams, E.D.: Diarrhea and thyroid cancer. Proc. R. Soc. Med. 59: 602 (1966).
536 Wiliams, E.D.: Histogenesis of medullary carcinoma of the thyroid. J. clin. Path. 19: 114–118 (1966).
537 Williams, E.D.; Brown, C.L.; Doniach, I.: Pathological and clinical findings in a series of 67 cases of medullary carcinoma of the thyroid. J. clin. Path. 19: 103–113 (1966).
538 Williams, E.D.; Pollock, D.J.: Multiple mucosal neuromata with endocrine tumors. A syndrome allied to von Recklinghausen's disease. J. Path. Bact. 91: 71–80 (1966).
539 Williams, E.D.: Medullary carcinoma of the thyroid. J. clin. Path. 20: 395–398 (1967).
540 Williams, E.D.; Karin, S.M.M.; Sandler, M.: Prostaglandin secretion by medullary carcinoma of the thyroid. A possible cause of the associated diarrhea. Lancet i: 22–23 (1968).
541 Williams, E.D.; Morales, A.M.; Horn, R.C.: Thyroid carcinoma and Cushing's syndrome. A report of two cases with a review of the common features of the 'non-endocrine' tumours associated with Cushing's syndrome. J. clin. Path. 21: 129–135 (1968).
542 Williams, E.D.: Medullary carcinoma of the thyroid; in Taylor, Calcitonin 1969, pp. 483–486 (Heinemann, London 1970).
543 Williams, E.D.: The pathology of thyroid malignancy. Br. J. Surg. 62: 757–759 (1975).
544 Williams, E.D.: Histogenesis of medullary carcinoma of the thyroid. J. clin. Path. 19: 114–118 (1966).
545 Williams, G.D.; Hargis, G.K.; Ensinck, J.W.; Kukreja, S.C.; Bowser, E.N.; Chertow, B.S.; Henderson, W.J.: Role of endogenous somatostatin in the secretion of parathyroid hormone and calcitonin. Metabolism 28: 950–954 (1979).
546 Wolfe, H.J.; Melvin, K.E.W.; Cervi-Skinner, S.J.; Al Saadi, A.A.; Juliar, J.F.; Jackson, C.E.; Tashjian, A.H., Jr.: C-cell hyperplasia preceding medullary thyroid carcinoma. New Engl. J. Med. 289: 437–441 (1973).
547 Wolfe, H.J.; Voelkel, E.F.; Tashjian, A.H., Jr.: Distribution of calcitonin-containing cells in the normal adult human thyroid gland. A correlation of morphology with peptide content. J. clin. Endocr. Metab. 38: 688–694 (1974).
548 Woodhouse, N.J.Y.; Reiner, M.; Kalu, D.N.; Galante, L.; Leese, B.; Foster, G.V.;

Joplin, G.F.; MacIntyre, I.: Some effects of acute and chronic calcitonin M administration in man; in Taylor, Calcitonin 1969, Proc. 2nd int. Symp., London 1969, pp. 504–513 (Heinemann, London 1970).
549 Woodhouse, N.J.Y.; Bordier, P.; Fisher, M.; Joplin, G.F.; Reiner, M.; Kalu, D.M.; Foster, G.V.; MacIntyre, I.: Human calcitonin in the treatment of Paget's bone disease. Lancet *i:* 1139–1143 (1971).
550 Woolner, L.B.; Beahrs, O.H.; Black, B.M.; McConahey, W.M.; Keating, F.R., Jr.: Classification and prognosis of thyroid carcinoma. Am. J. Surg. *102:* 354–387 (1961).
551 Wurtman, R.J.: Control of epinephrine synthesis in the adrenal medulla by the adrenal cortex. Hormonal specificity and dose-response characteristics. Endocrinology *79:* 608–614 (1966).
552 Yamada, Y.; Ito, S.; Matsubara, Y.; Kobayashi, S.: Immunohistochemical demonstration of somatostatin-containing cells in the human, dog, and rat thyroids. Tohoku J. exp. Med. *122:* 87–92 (1977).
553 Zeytinoglu, F.U.; Gagel, R.F.; Tashjian, A.H., Jr.; Hammer, R.A.; Leeman, S.E.: Characterization of neurotensin production by a line of rat medullary thyroid carcinoma cells. Proc. natn. Acad. Sci. USA *77:* 3741–3745 (1980).

Subject Index

ACTH 38, 50
Adrenal medullary hyperplasia 59, 60
Aminoguanidine 38
Amyloid 12, 13
Athyreotic cretins 22
Auerbach's plexus 71

Beta-endorphin 41, 50
Biological effects of calcitonin, see
   Calcitonin, biological effects
Bone 43
Bone disease 34
Bovine calcitonin 18
Bradykinin 41

C cells 1, 14
   Adenoma 15
   Carcinoma, see Medullary thyroid
     carcinoma
   Hyperplasia 14
Café-au-lait spots 71
Calcification 13, 74
Calcitonin
   Biochemistry 18
   Biological effects 43
     Adenylate cyclase 22
     Bone 43
     Gastrointestinal 43
     Kidney 43
     Mineral metabolism 44
     Neurotransmitter 46
   Clinical uses 29
   Immunochemical heterogeneity 27, 28
   Metabolism 28
   Precursors 27

Calcitonin secretion
   Age and sex 22
   Athyreotic cretins 22
   Autoregulation 22
   Bone disease 24
   by MTC 23
   by normal C cells 20
   Ectopic 31
   Eutopic 30
   Hypercalcemia 32
   Hypocalcemia 33
   Iodine 22
   Malignancy 30
   Pancreatitis 34
   Primary hyperparathyroidism 32
   Prostaglandins 22
   Provocative testing 24
     Calcium 24
     Gastrointestinal factors 21
     Glucagon 24
     Magnesium
     Neuroendocrine factors 21
     Pentagastrin 24
     Pentagastrin vs. calcium 25
     Tetragastrin 25
     Whiskey 26
   Renal disease 33
   Strontium 22
Carcinoid syndrome 52, 54
C-cell adenoma 15
C-cell hyperplasia 14
CEA 41
Cervical enlargement 54
Cushing's syndrome 38, 50, 54
Cutaneous lesions 69

Subject Index

Diabetes mellitus 42
Diagnosis
   Hypercalcitoninemia 30
   Medullary thyroid carcinoma 6, 8
   Pheochromocytoma 57, 58
   Venous catheterization procedures 26
Diarrhea 39, 50–52
Diverticulosis 71

Ectopic calcitonin secretion 30, 31
Erythremia 42

Gastrointestinal abnormalities 43, 71
Genetics
   Medullary thyroid carcinoma 7
   Multiple endocrine neoplasia 55
Glucagon 74
Gynecomastia 42

Hirschsprung's disease 71
Histaminase 38
History
   ACTH and Cushing's syndrome 36
   Hyperparathyroidism 60
   MTC 1
   Multiple mucosal neuroma syndrome 65
   Pheochromocytoma 56
Hypercalcemia 33
Hyperparathyroidism 55, 60
   History 60
   Incidence 61
   Pathology 61
   Relationship to MTC 64
   Treatment 78
Hypocalcemia 33

Immunochemical heterogeneity 26–28, 32, 33
Immunohistochemistry 2, 14
Incidence
   Hyperparathyroidism 61
   Medullary thyroid carcinoma 5
   Pheochromocytoma 57
Intestinal ganglioneuromatosis 66, 71

Kidney stones 43

Lactation 22, 46

Magnesium 26
Marfanoid habitus 53, 55, 67–69, 72
   Muskuloskeletal abnormalities in 72
Medullary thyroid carcinoma
   Animal tumors 2
   Calcitonin secretion 18
   Clinical manifestations 47
      Bone disease 48
      Carcinoid syndrome 52
      Cervical enlargement 54
      Cushing's syndrome 54
      Diarrhea 50
      Gynecomastia 53
      Hypertention 53
      Kidney stones 50
      Marfanoid habitus 53
      Minerals 49
      Mucosal neuromas 54
      Peptic ulcer disease 51
      Pigmentation 53
   Genetics 7
   Histogenesis 4
   Histology 9
   History 1
   Incidence 5
   Natural history 16
   Pathology 8
   Radiological manifestations 13, 74, 75
   Secretory products
      ACTH 36
      Beta-endorphin 41
      Calcitonin 18
      CEA 41
      Dopa decarboxylase 40
      Histaminase 38
      MSH 41
      Nerve-growth factor 41
      Neurotensin 41
      Prostaglandins 39
      Releasing factors 41
      Somatostatin 41
      Substance P 41
   Treatment 76
Megacolon 71, 72
Meissner's plexus 71

Subject Index

MEN, *see* Multiple endocrine neoplasia
Metabolism of calcitonin 28
Mineral metabolism 44, 45
MMN, *see* Multiple mucosal neuromas
MSH 41, 50
MTC, *see* Medullary thyroid carcinoma
Mucosal neuroma syndrome 65
    Clinical features 66
    Gastrointestinal abnormalities 71
    History 65
    Marfanoid habitus 72
Mucosal neuromas 68
    Buccal 68
    Cutaneous 69
    Lips 68
    Ocular 68
    Oral 68, 73
    Tongue 68
    Treatment 78
Multiple endocrine neoplasia 55
    Classification 56
    Hyperparathyroidism 55, 60
    Mucosal neuromas 65
    Pheochromocytoma 56
    Radiological findings 74
    Treatment 76
Multiple mucosal neuromas 65

Nerve-growth factor 41, 50
Neural crest 3, 4, 11, 40, 52, 55
Neuroendocrine 21, 46
Neurotensin 41
Neurotransmitter 21, 46
Nutmeg 51

Oat cell carcinoma of the lung 11, 30
Ocular abnormalities 68
Osteoporosis 32–34
Ovine calcitonin 18

Pancreatitis 34
Parafollicular cells, *see* C cells
Parathyroid gland 61, 63
Parathyroid hormone secretion in MTC 63
Pathogenesis
    ACTH and Cushing's syndrome 36

Hyperparathyroidism 64
Multiple endocrine neoplasia 37
Pathology
    Hyperparathyroidism 61
    Medullary thyroid carcinoma 2
    Multiple mucosal neuroma syndrome 67
    Pheochromocytoma 58, 59
Pentagastrin 24
Peptic ulcer 51, 42
Pheochromocytoma 55, 56, 58
    Adrenal medullary hyperplasia 59
    Clinical manifestations 57
    History 56
    in MEN 56
    Treatment 77
Pituitary 46
Polycythemia vera 42
Porcine calcitonin 18
Precursors of calcitonin 27
Pregnancy 22, 46
Prognathism 73
Prostaglandins 39, 40, 53
Provocative testing of calcitonin secretion 24

Radiological manifestations
    Calcification 74
    Medullary thyroid carcinoma 13, 74, 75
    Megacolon 75
    Megaloureter 75
    Pheochromocytoma 75
Releasing factors 42
Renal disease 33

Salmon calcitonin 18
Scheuermann's disease 72
Somatostatin 41
Substance P 41

Tetragastrin 25
Treatment
    Hypercalcemia 29
    Hyperparathyroidism 78
    Medullary thyroid carcinoma 76
    Mucosal neuromas 78

Treatment (cont.)
  Multiple endocrine neoplasia 76
  Paget's disease 29
  Pheochromocytoma 77
  Serial calcitonin measurements 26
  Venous catheterization 26
  with calcitonin 29

UB gland, *see* Ultimobranchial gland
Ultimobranchial gland 2, 18

Vasoactive intestinal polypeptide 42
Venous catheterization procedures 25, 26
Vitamin D 49

Whiskey 26